Motor Vehicle Mechanic's Textbook

Fifth Edition

F. K. Sully

Heinemann Professional Publishing

Heinemann Professional Publishing Ltd
Halley Court, Jordan Hill, Oxford OX2 8EJ

OXFORD   LONDON   MELBOURNE   AUCKLAND

First published by Newnes-Butterworths 1957
Second edition 1959
Reprinted 1960, 1962, 1963, 1965
Third edition 1968
Reprinted 1970, 1975
Fourth edition 1979
Reprinted 1980, 1982
Fifth edition published by Heinemann Professional Publishing 1988

© F. K. Sully 1988

British Library Cataloguing in Publication Data
Sully, F. K.
  Motor vehicle mechanic's textbook. –
  5th ed.
  1. Motor vehicles
  I. Title
  629.2   TL145

ISBN 0 434 91884 9

Typeset in Great Britain by Keyset Composition, Colchester
Printed in Great Britain by L. R. Printing Services, Crawley

Contents

Preface
1	Calculations and science	1
2	Drawing	27
3	The vehicle structure	31
4	Suspension	36
5	The front-wheel assembly	46
6	Engine principles	68
7	Engine components	89
8	The lubrication system	110
9	The cooling system	122
10	The fuel system	132
11	Carburation	140
12	The compression-ignition engine	160
13	The electrical system	180
14	The ignition system	186
15	The charging and starting systems	195
16	The clutch	207
17	The gearbox	216
18	Automatic transmission	235
19	Universal joints: propeller and drive shafts	241
20	The final drive and differential	247
21	The rear-wheel assembly	258
22	The braking system	269
23	Materials	298
24	Safety in the motor vehicle workshop	305

Appendix Conversion factors 310

Index 312

Preface to the Fifth Edition

Present-day production of motor vehicles in all parts of the world necessitates the availability of fully trained service mechanics if these vehicles are to be maintained in a state of efficiency and ensure a viable capital investment for the company or private owner. The development of the natural resources of many home and overseas areas relies largely on the satisfactory operation of commercial and public service vehicles, and this, in turn, is entirely dependent on correct maintenance and repair.

In recent years the replacement of complete assemblies – rather than the repair of individual items from the unit – has become the mechanic's standard procedure. With the increasing application of the microcomputer to motor vehicle control systems, fault diagnosis has necessarily become more sophisticated. However, in spite of such changes, a simple and clear understanding of basic principles remains fundamental to satisfactory work and to job satisfaction.

Motor Vehicle Mechanic's Textbook covers various City and Guilds course syllabuses. Special attention has been paid to preserving the balance between theory and practice, a sound knowledge of both being essential to the art of diagnosis. The book will also prove useful to those engaged in the maintenance, repair and overhaul sections of the motor industry, as well as to motorists who wish to know more about their vehicles.

For this Fifth Edition the text has been substantially revised and is illustrated by over 200 line illustrations. The SI system of units is employed throughout and, as a few non-SI units remain in motor vehicle usage, a comprehensive conversion table is included.

F. K. Sully

Chapter 1
Calculations and science

1.1 International system of units (SI)

The Système Internationale d'Unités was adopted in 1960 as the title for an MKSA system based on the metre (m), the unit of length; the kilogram (kg), the unit of mass; the second (s), the unit of time; the ampere (A), the unit of electric current; the kelvin (K), the degree of temperature; and the candela (cd), the unit of light intensity.

Associated with these basic units are a variety of supplementary derived units which are adopted worldwide.

Derived units

Physical quantity	SI unit	Unit symbol
Force	newton	$N = kg\,m/s^2$
Work, energy, quantity of heat	joule	$J = N\,m$
Power	watt	$W = J/s$
Electric charge	coulomb	$C = A\,s$
Electrical potential	volt	$V = W/A$
Electric capacitance	farad	$F = A\,s/V$
Electric resistance	ohm	$\Omega = V/A$
Frequency	hertz	$Hz = s^{-1}$

Multiplying factors

Factor	Prefix	Symbol
10^6	mega	M
10^3	kilo	k
10^2	hecto	h
10	deca	da
10^{-1}	deci	d
10^{-2}	centi	c
10^{-3}	milli	m
10^{-6}	micro	μ

1.2 Mensuration

Area

Square	l^2	where:	l = length of side
Rectangle	lb		b = breadth
Parallelogram	lh		h = perpendicular height
Triangle	$\frac{1}{2}lh$		r = radius (smallest)
Circle	πr^2		R = radius (largest)
Annulus	$\pi(R+r)(R-r)$		$\pi = \dfrac{\text{circumference}}{\text{diameter}} = 3.1416$

The perimeter of a circle is $2\pi r$ or πd, where $d = 2r$ is the diameter.

Solids

Solid	Volume	Total surface area
Cube	l^3	$6l^2$
Square prism	lbh	$2(lb + bh + hl)$
Cylinder	$\pi r^2 h$	$2\pi r(r + h)$
Cone (slant height l)	$\frac{1}{3}\pi r^2 h$	$\pi r(r + l)$
Sphere	$\frac{4}{3}\pi r^3$	$4\pi r^2$

Figure 1.1 shows a circle, a cylinder and a cone.

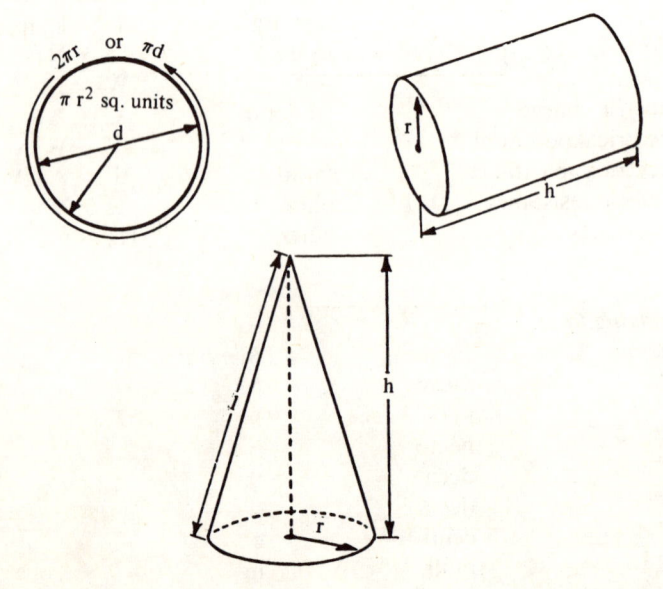

Figure 1.1   Circle, cylinder, cone

1.3 Geometry

In any triangle the sum of the three angles is two right angles, 180°. The longest side is opposite the largest angle and the shortest side opposite the smallest angle.

In a right-angled triangle, the sum of the other two – complementary – angles is 90°. The longest side, opposite the right angle, is called the *hypotenuse*.

In any right-angled triangle the square on the hypotenuse is equal to the sum of the squares on the other two sides. This – Pythagoras' – theorem is useful when checking or setting out 90°. A triangle can be formed whose sides are in the proportion 3:4:5 or 5:12:13; either will ensure a right angle opposite the longest side (Figure 1.2).

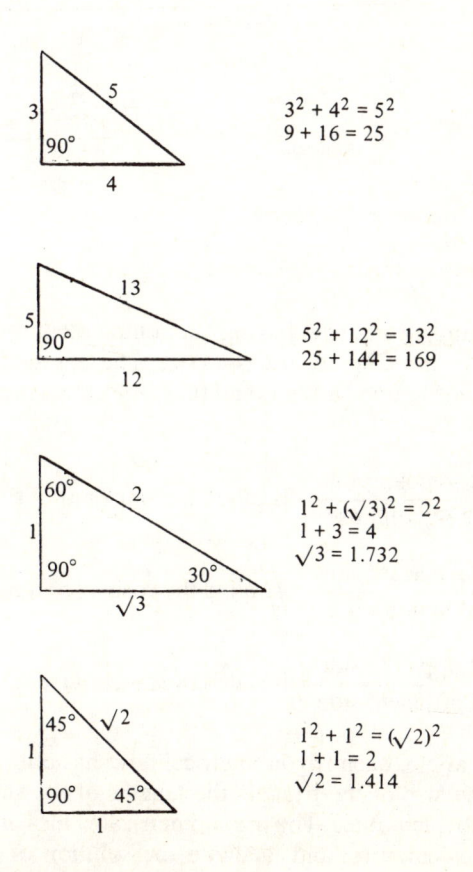

$3^2 + 4^2 = 5^2$
$9 + 16 = 25$

$5^2 + 12^2 = 13^2$
$25 + 144 = 169$

$1^2 + (\sqrt{3})^2 = 2^2$
$1 + 3 = 4$
$\sqrt{3} = 1.732$

$1^2 + 1^2 = (\sqrt{2})^2$
$1 + 1 = 2$
$\sqrt{2} = 1.414$

Figure 1.2 Pythagoras' theorem

1.4 Trigonometry

In a right-angled triangle, given one of the complementary angles, the side opposite the given angle is called the *opposite* side and the side nearest the given angle is called the *adjacent* side; the remaining side is the hypotenuse (Figure 1.3).

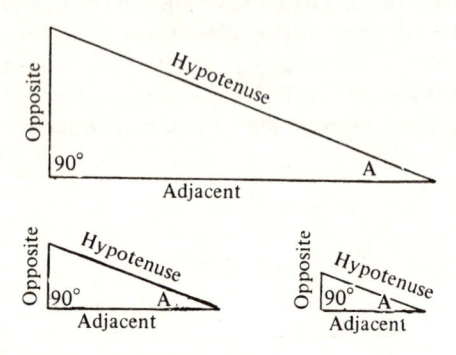

Figure 1.3 Trigonometric ratios

All right-angled triangles having one complementary angle of a given size are similar in shape, regardless of size. The lengths of their sides bear the same ratios to one another, and these are called the trigonometrical ratios. Thus:

$$\frac{\text{length of opposite side}}{\text{length of hypotenuse}}$$ is called the sine (sin) of the angle

$$\frac{\text{length of adjacent side}}{\text{length of hypotenuse}}$$ is called the cosine (cos) of the angle

$$\frac{\text{length of opposite side}}{\text{length of adjacent side}}$$ is called the tangent (tan) of the angle

For a given angle, each trigonometrical ratio has only one value, since whatever the size of the triangle the lengths of the sides will bear the same ratio to each other. The trigonometrical values are available from tables and calculators, and facilitate the solution of some workshop problems.

Calculations and science

1.5 Mass and weight

A 'body' contains a certain amount of 'stuff' or matter called its *mass*. The unit of mass is the kilogram (kg).

The pull of the earth – the force of gravity – acting on this mass is the *weight* of the body.

Owing to its mass a body has *inertia* – that is, it resists being accelerated or decelerated and will remain at rest or continue moving at a uniform speed in a straight line unless acted upon by an external force.

1.6 Density

Density is the mass of a substance per unit volume (kg/m^3). The density of water is, for practical purposes, 1000 kg/m^3 or 1 kg/l. (The litre (l) is 10^{-3} m^3.)

1.7 Relative density or specific gravity

The ratio

$$\frac{\text{mass of a substance}}{\text{mass of an equal volume of water}}$$

Table 1.1

Substance	Relative density
Oxygen	0.0014
Cork	0.22
Paraffin/petrol	0.7/0.8
Water	1.0
Magnesium	1.7
Carbon	2.0
Glass	2.6
Aluminium	2.7
Chromium	6.6
Tin	7.3
Iron/steel	7.2/8.0
Copper	8.3
Nickel	8.9
Molybdenum	10.0
Lead	11.4
Mercury	13.6
Platinum	21.5

In many cases only approximate figures can be given

is called the relative density of the substance, and represents how many times it is heavier or lighter than the same volume of water (Table 1.1). Note that relative density has no units.

1.8 Speed and acceleration

Speed is measured as metres per second (m/s), or sometimes more conveniently as kilometres per hour (km/h). Useful conversions are 5 m/s = 18 km/h and 0.278 m/s = 1 km/h.

Acceleration or deceleration is the rate of change of speed. It is measured as metres per second per second or m/s^2.

An increase in speed from 36 km/h to 72 km/h during 4 s (that is from 10 m/s to 20 m/s) is an average acceleration of 2.5 m/s every second or 2.5 m/s^2.

If the speed increases from u m/s to v m/s during t seconds, then the average acceleration a m/s^2 is given by

$$a = \frac{v - u}{t} \text{ m/s}^2$$

1.9 Acceleration due to gravity

In a vacuum all freely falling bodies, whatever their size, shape or mass, have the same acceleration at a given place (Figure 1.4). This acceleration, given the symbol g since it is due to the force of gravity, has the value of about 9.81 m/s^2 at sea level near London, 9.78 m/s^2 at the equator and 9.83 m/s^2 at the poles.

The acceleration of objects falling in the atmosphere depends on their wind resistance. For example, depending on the conditions, the human body reaches a terminal velocity of some 200 km/h, when the wind resistance equals the force of gravity and no further acceleration can occur. A motor vehicle is also subject to wind resistance; a typical speed–time graph for a vehicle is shown in Figure 1.4.

1.10 Force

A force may be simply described as a push or a pull on a body – an action which tends to move a body from rest or alter its speed or direction of movement.

Force is measured in newtons (N); 1 newton is the force needed to give a mass of 1 kg an acceleration of 1 m/s^2. The acceleration of 1 kg due to gravity is about 9.81 m/s^2; hence the weight of 1 kg is about 9.81 N.

Calculations and science

Average speed during each second (in m/s) equals distance covered in each second (in m)	Scale position of body time in seconds	Speed at end of each second (m/s)	Total distance covered at end of each second (m)
$\frac{0 + 9.8}{2} = 4.9$	0	0	0
	1	9.8	4.9
$\frac{9.8 + 19.6}{2} = 14.7$			
	2	19.6	19.6
$\frac{19.6 + 29.4}{2} = 24.5$			
	3	29.4	44.1
$\frac{29.4 + 39.2}{2} = 34.3$			
	4	39.2	78.4
$\frac{39.2 + 49.0}{2} = 44.1$			
	5	49.0	122.5

Relationship of distance and speed for a freely falling body (taking $g = 9.8$ m/s^2)

Distance-time graph of freely falling body

Speed-time graph for freely falling body

Acceleration-time graph for freely falling body

Speed-time graph for motor vehicle, acceleration not constant

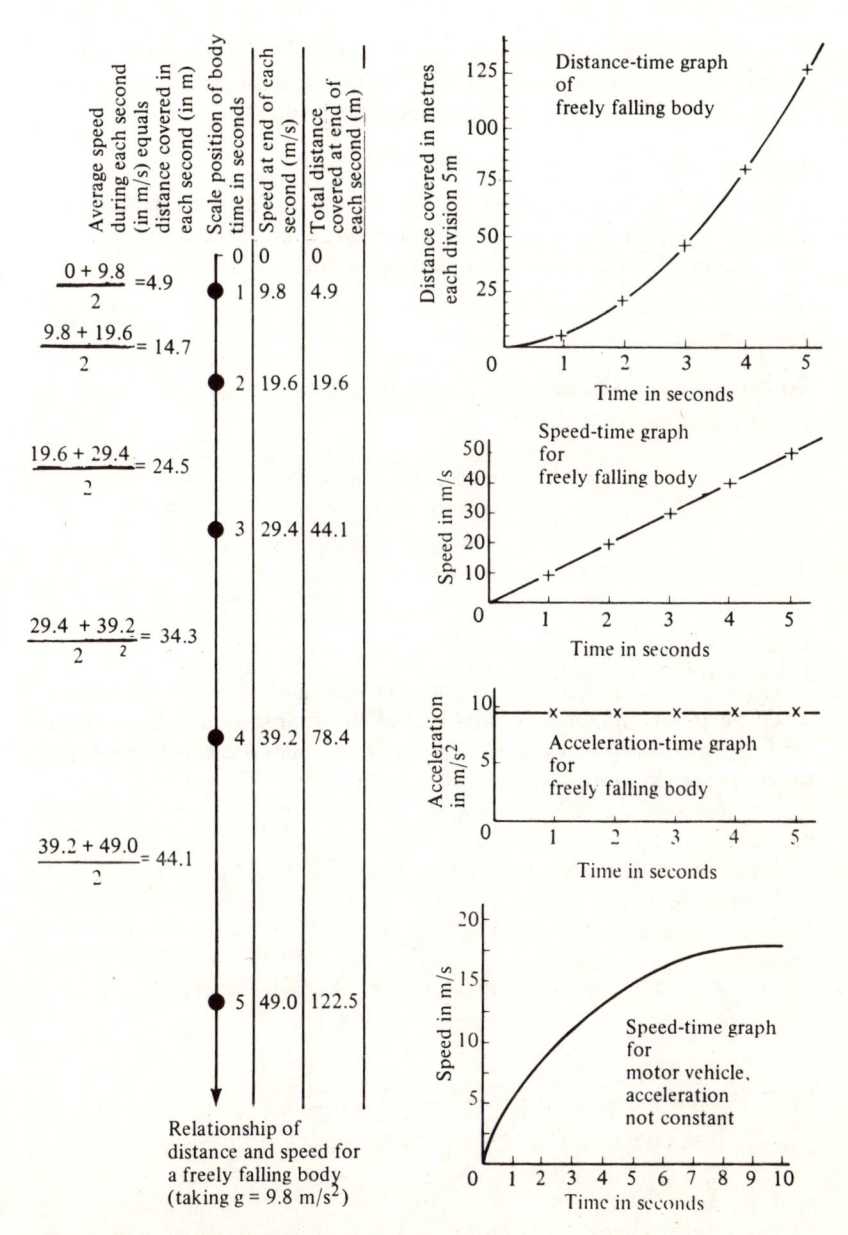

Figure 1.4 Freely falling body in a vacuum

1.11 Work

Work is done when a force overcomes resistance and causes movement. Work is measured by the product of the force and the distance moved in the direction of the force, the unit being the joule (J):

$$W = Fs$$

where W = work done in joules (J), F = force in newtons (N), and s = distance in metres (m) moved in the direction of the force.

If the force causes no movement, then no work is done.

1.12 Power

Power is the rate of doing work. The unit, the watt, is a rate of working of 1 joule per second (1 J/s):

$$\begin{aligned}\text{power} &= \text{work done per second} \\ &= \text{newton} \times \text{metre per second} \\ &= \text{joule per second} \\ &= \text{watt}\end{aligned}$$

1.13 Torque

When a force acts on a body pivoted on a fixed axis, the product of the force perpendicular to the radius, and the radius at which it acts, is termed the *turning moment* of the force or *torque*. Torque is measured in newton metres (Nm) (to distinguish it from work). From Figure 1.5:

$$T = Fr$$

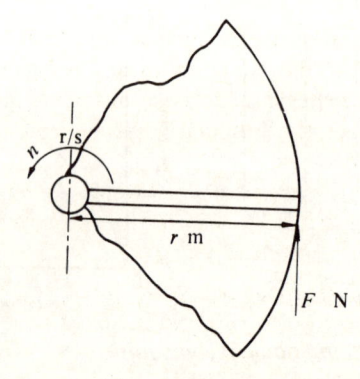

Figure 1.5 Torque

where F = force in newtons (N), r = radius in metres (m), and T = torque in newton metres (Nm).

1.14 Work done by torque

The work done by the torque per revolution is the product of the force and the distance moved in the direction of the force – that is, the circumference:

$$W = F \times 2\pi r$$
$$= 2\pi F r$$

where W = work done in joules (J), F = force in newtons (N), r = radius in metres (m) at which the force acts.

The work done in n revolutions will be:

$$W = 2\pi F r n \quad \text{joules}$$

If n revolutions are made per second, then the work done per second is $2\pi F r n$ joules. This is the *power* produced:

$$P = 2\pi F r n$$

or

$$P = 2\pi T n$$

where P = power in watts (W), Fr or T = torque in joules (J), and n = rotational speed (rev/s).

Using these formulae, the power can be calculated from the torque and speed of a shaft.

1.15 Principle of moments

When a body is at rest or in equilibrium (a state of balance), the sum of the clockwise turning moments about any axis, real or imaginary, is equal to the sum of the anticlockwise moments about the same axis. Were this not the case, the unbalanced moment would cause the body to rotate about the chosen axis.

As an example, a beam of negligible weight has loads as shown in Figure 1.6 and is pivoted at P so as to be in equilibrium. Taking moments about P:

sum of anticlockwise moments = sum of clockwise moments
$10 \times (0.08 + 0.06) + 12 \times 0.06 = 4 \times 0.53$
$1.40 + 0.72 = 2.12$
$| 2.12 \, \text{Nm} = 2.12 \, \text{Nm}$

The pivot P must exert an upward force on the beam equal to the sum of the downward forces of 10 N + 12 N + 4 N. This upward force of 26 N has no turning moment in this case since it is acting through the axis.

If we take moments about an imaginary axis at B, then:

$$\text{sum of anticlockwise moments} = \text{sum of clockwise moments}$$
$$10 \times 0.08 + 26 \times 0.06 = 4 \times (0.53 + 0.06)$$
$$0.8 + 1.56 = 4 \times 0.59$$
$$2.36 \text{ Nm} = 2.36 \text{ Nm}$$

The force of 12 N exerts no turning moment in this case as it is acting through the chosen axis.

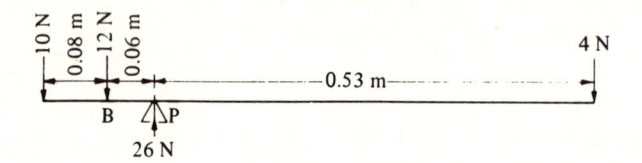

Figure 1.6

1.16 Centre of gravity

The centre of gravity (c. of g.) of a body can be regarded as the point where, if the whole weight of the body were concentrated, it would produce a moment of force about any axis equal to the sum of the moments of force of each part of the body about the same axis. When inertia force is involved the centre of gravity becomes the centre of mass.

As an example, consider a body consisting of weights of 10 N, 12 N and 4 N located on a beam of negligible weight, as shown in Figure 1.7. Let L be the distance of the c. of g. of the body from an axis, say 0.01 m from the right-hand end of the beam. Taking moments about that axis:

total weight of body (concentrated at c. of g.)
× distance of c. of g. from any axis = sum of the moments of each part of the body about the same axis

$$(10 + 12 + 4) \times L = 4 \times 0.01 + 12 \times 0.60 + 10 \times 0.68$$
$$26 \times L = 0.04 + 7.20 + 6.80$$

$$L = \frac{14.04}{26}$$

$$= 0.54 \text{ m}$$

Calculations and science

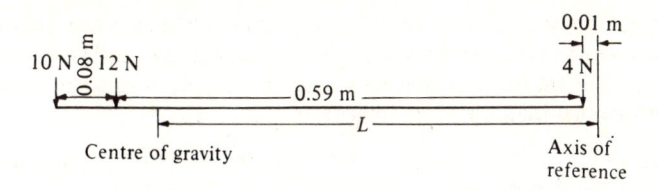

```
              0.08 m                              0.01 m
    10 N  ┤├ 12 N                                  4 N
    ├─┤                 0.59 m
    ├────────────────────── L ──────────────────────┤
        Centre of gravity                    Axis of
                                             reference
```

Figure 1.7

The c. of g. is 0.06 m from the 12 N weight and 0.53 m from the 4 N weight, and this is the pivot point about which the body would balance.

1.17 Couple

When two equal forces act on a body so that their lines of action are parallel but opposite in direction, they form a couple tending to rotate the body.

The torque produced by a force acting on a pivoted body can be regarded as the result of a couple formed by the original force and an equal and opposite *reaction* at the pivot (Figure 1.8).

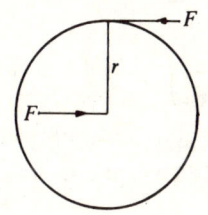

Figure 1.8 Couple

A couple can only be balanced by another couple of equal value acting in the opposite direction of rotation and not by a single force.

1.18 Inertia force

All bodies have inertia – the tendency to remain at rest or in uniform motion. For example, when a piston is decelerated from maximum speed to a dead centre position and accelerated in the opposite direction, it exerts an inertia force on the connecting rod.

The value of the inertia force depends upon the mass of the body and the acceleration or deceleration:

$$F = ma$$

11

where F = force in newtons (N), m = mass in kilograms (kg), and a = acceleration or deceleration in metres/second/second (m/s^2).

Thus to reduce the inertia forces produced by the reciprocating parts, their mass must be kept as small as possible.

1.19 Centrifugal force

A moving body travels in a straight line at uniform speed unless acted upon by an external force. If made to travel in a circle, the body exerts *centrifugal* force acting outwards from the centre upon the constraining member. The equal and opposite constraining force is termed *centripetal*.

$$CF = \frac{mv^2}{r}$$
$$= m\omega^2 r$$

where CF = centrifugal force in newtons (N), m = mass in kilograms (kg), r = radius in metres (m), v = linear velocity in metres/second (m/s), and ω = angular velocity in radians/second (rad/s), where 1 revolution = 2π radians.

1.20 Mean piston speed

The product of twice the stroke, measured in metres, and the rotational speed (rev/s) of the engine gives the mean or average piston speed in m/s.

The higher the mean piston speed, the greater are the inertia forces of the reciprocating parts. The maximum mean piston speed usually employed with current production engines is about 16 m/s.

1.21 Friction force

It is found that the horizontal force required to drag a body over a smooth, level, dry surface is approximately a constant fraction of the perpendicular force between the surfaces. In Figure 1.9, W is the load or

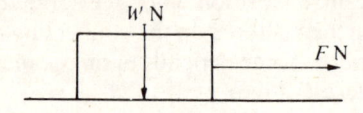

Figure 1.9 Friction

normal force between the surfaces, and F is the force parallel to the surfaces or the friction force. The ratio F/W, fairly constant for any given combination of two materials, is termed the *coefficient of friction* μ. The value of μ depends upon the materials and the condition of the surfaces and not greatly upon the speed or area of contact.

In some cases – brakes, drive belts, clutch, tyres – a high coefficient of friction is required, whilst in others, such as bearings, a low coefficient is necessary.

The friction force reaches a maximum value just before sliding occurs; this is static friction or *stiction*. It then reduces to a lower dynamic or kinetic value during sliding. The difference between these two values can have important results, e.g. during a cornering 'breakaway' when the centrifugal force exceeds the limit of the static friction force.

The coefficient of friction between the tyres and the road may be almost zero under icy conditions. Alternatively, some braking distances indicate a coefficient exceeding 1.0, i.e. the frictional force is greater than the weight of the vehicle. Some interlocking between the tyres and the road takes place where tread rubber is torn away.

When surfaces are separated by a film of lubricant, fluid friction rather than dry friction is involved. This viscous friction is related to the area of contact, the speed and the viscosity of the lubricant.

Materials	Coefficient of static friction
Leather on cast iron	0.4 –0.6
Friction material on cast iron	0.35 –0.45
Cast iron on cast iron	0.2 –0.3
Smooth greased wooden 'skids'	0.05 –0.08
Plain bearing fluid friction	0.02 –0.04
Ball bearing	0.002–0.004

1.22 Pressure

Pressure is the force per unit area; the unit is N/m^2 or the pascal (Pa). Larger practical units are kN/m^2 (kPa) and MN/m^2 (MPa). Note that $1\ MN/m^2 = 1\ N/mm^2$.

A pressure of 7 MPa means that each mm^2 subject to the pressure has a force of 7 N acting on it, and the total force on the surface will be the product of the pressure and the area:

$$\text{pressure} = \frac{\text{force}}{\text{area}}$$

$$\text{force} = \text{pressure} \times \text{area}$$

1.23 Atmospheric pressure

Air has weight. The atmosphere above the earth produces a pressure at sea level of approximately 1 bar, where 1 bar = 10^5 N/m^2 or 10^5 Pa. Standard atmospheric pressure is 1.013 25 bar, 1013.25 millibar or 101.325 kPa. Above sea level the atmospheric pressure will be less than 1 bar.

1.24 Gauge and absolute pressure

The ordinary pressure gauge gives readings measured above atmospheric pressure. To obtain the absolute pressure, that is the pressure measured above a perfect vacuum, atmospheric pressure must be added to the gauge reading:

absolute pressure = gauge pressure + atmospheric pressure

1.25 Stress

Components may be subject to a *tensile* load, as in a brake cable; a *compressive* load, as in a push rod; or a *shear* load, as in a shackle pin. In each case the value of the stress produced is found by dividing the load by the cross-sectional area on which it acts. Typical units of stress are N/m^2 (Pa), kN/m^2 (kPa) and MN/m^2 (MPa).

For tensile and compressive stress the cross-sectional area is measured at right angles to the direction of the force, while for shear stress the area is measured parallel to the direction of the force.

1.26 Temperature

The temperature, or degree of 'hotness' of a body, can be measured on the degree Celsius (°C) or the kelvin (K) scale. The upper 'fixed' point – that is, the boiling point of water at atmospheric pressure – is 100°C or 373.15 K, and the lower fixed point, the freezing point of water, is 0°C or 273 K.

The melting points of various substances are shown in Table 1.2.

1.27 Thermodynamic scale of temperature

At constant pressure the volume of a given mass of gas increases or decreases by (for practical purposes) 1/273 of its volume at 0°C for each degree rise or fall. A theoretically 'perfect' gas would have no volume and no internal energy at −273°C.

The thermodynamic temperature scale uses −273°C as zero and

Table 1.2

Substance	Melting point (°C)
Oxygen	−219
Mercury	−39
Tin	232
Lead	327
Zinc	419
Magnesium	633
Aluminium	658
Brass	900
Copper	1083
Glass	1100
Iron (cast)	1200
Iron (wrought)	1530
Nickel	1452
Chromium	1520
Platinum	1755
Molybdenum	2450
Carbon	3500

temperature intervals in kelvin, where 1 kelvin exactly corresponds to 1 degree Celsius:

$$T = t + 273$$

where T = temperature in kelvin (K) and t = temperature in degrees Celsius (°C).

1.28 General gas law

When the temperature, or volume, or pressure of a given mass of gas changes, the relationship between them before and after the change is given by

$$\frac{p_1 V_1}{T_1} = \frac{p_2 V_2}{T_2}$$

where
  p_1 = initial absolute pressure
  V_1 = initial volume
  T_1 = initial thermodynamic temperature
  p_2 = final absolute temperature
  V_2 = final volume (in the same units as V_1)
  T_2 = final thermodynamic temperature

1.29 Unit of heat

Heat is one of many forms of energy. It is stored in a substance in the kinetic (movement) energy of the molecules.

The unit of heat is the joule (J), which is also the unit of work. When mechanical energy is converted into heat energy the quantity of work done equals the quantity of heat produced, both measured in joules. This relationship is known as the *first law of thermodynamics*.

In practice it is often convenient to measure heat in kilojoules (kJ).

1.30 Specific heat capacity

The quantity of heat required to raise a mass of 1 kg of a substance through 1°C or 1 K is called the specific heat capacity of the substance. The units are joules per kilogram per degree Celsius (J/kg°C).

Table 1.3

Substance	Specific heat capacity (J/kg°C)
Lead	130
Mercury	130
Tin	225
Copper	390
Iron	500
Glass	600
Aluminium	900
Rubber	1500
Petrol	1800
Ice	2100
Paraffin wax	2900
Water	4190

The specific heat capacities of various substances are shown in Table 1.3. Approximate figures only can be given in some cases, since the specific heat capacity depends upon the composition of the substance and the temperature range involved.

1.31 Quantity of heat

The quantity of heat required to raise a given mass of any substance through a given range of temperature can be calculated from

$$Q = mct$$

where Q = quantity of heat (J), m = mass of substance (kg), c = specific heat capacity of substance (J/kg°C), and t = rise in temperature (°C or K).

In a cooling process, t is the fall in temperature and Q is the quantity of heat lost by the body.

1.32 Specific latent heat

Heat is required to change a substance from the solid to the liquid state and from the liquid to the gaseous state. During a change of state the temperature remains constant at the melting or boiling point of the substance and the heat required is termed *latent heat*. During the reverse change of state, from gas to liquid or liquid to solid, latent heat is given out by the substance.

The heat required to change 1 kg of a substance from the solid to the liquid state is called the specific latent heat of *fusion* of the substance, with units kJ/kg. Similarly the specific heat of *vaporization* is the heat needed to change 1 kg of the substance from the liquid state to vapour. At standard atmospheric pressure:

specific latent heat of fusion of ice = 335 kJ/kg

specific latent heat of vaporization of water = 2257 kJ/kg

These quantities of heat, required to change the state of the substance from ice to water or from water to steam without rise in temperature, may be compared with the sensible heat required to raise the temperature of water from 0°C to 100°C, which is 419 kJ/kg.

1.33 Transfer of heat

Heat may be transferred in three ways:

Conduction occurs when the heat energy passes from one particle of the substance to the next. Most metals are good conductors of heat; copper and aluminium are two of the best. Plastics, wood, cork and rubber are among the poor conductors of heat; they are employed as heat insulators. Liquids and gases are generally very poor conductors of heat.
Convection currents are produced in liquids and gases. One portion of the fluid is heated, expands and, becoming less dense, is displaced by the cooler and denser surrounding fluid.
Radiation is the transmission of heat energy from the surface of a body in the form of rays. A dull black surface will radiate and absorb heat more readily than a polished light surface.

Table 1.4

Substance	Thermal conductivity (W/m°C)
Silver	418.7
Copper	382.3
Gold	295.2
Aluminium	203.5
Zinc	110.1
Brass	108.9
Iron (cast)	75.37
Tin	62.81
Lead	34.75
Mercury	7.955
Glass	0.921
Carbon (graphite)	0.628
Water	0.590
Perspex	0.209
Rubber	0.188
Wood (oak)	0.147
Cork	0.050
Air	0.024

Examples of the transfer of heat in the engine are the conduction of heat through the cylinder walls, setting up convection currents in the coolant; and the radiation of heat from the exhaust manifold surface.

The thermal conductivities of various substances are shown in Table 1.4.

1.34 Conversion of energy

Energy can be converted from one form to another, e.g.:

(a) Chemical energy into heat energy by combustion
(b) Heat energy into mechanical energy by a heat engine – one form being the internal combustion engine
(c) Mechanical energy into electrical energy by a dynamo
(d) Electrical energy into mechanical energy by an electric motor
(e) Mechanical energy into heat energy by friction.

1.35 Coefficient of linear expansion

When a body expands, the fraction

$$\frac{\text{increase in length produced by 1°C temperature rise}}{\text{original length (same units as extension)}}$$

is termed the coefficient of linear expansion per °C.

Calculations and science

Table 1.5

Substance	Coefficient of linear expansion per °C
Invar (steel with 36% nickel)	0.000 000 87
Glass/platinum	0.000 009
Iron	0.000 011
Nickel	0.000 013
Copper	0.000 017
Brass	0.000 019
Tin	0.000 022
Aluminium	0.000 023
Zinc	0.000 026
Lead	0.000 029

The coefficients of linear expansion for various substances are shown in Table 1.5.

Example 1

A flywheel has a diameter of 356 mm. Measured on the rim of the flywheel the inlet valve opens 45 mm before top dead centre (TDC) and the exhaust valve closes 50 mm after TDC. Convert these measurements to degrees and state the valve overlap.

$$\begin{aligned}
\text{diameter of flywheel} &= 356 \text{ mm} \\
\text{circumference of flywheel} &= 356 \times \pi \text{ mm} \\
&= 1118.4 \text{ mm}
\end{aligned}$$

Now 1118 mm subtends an angle at flywheel centre of 360°. Therefore

$$\begin{aligned}
1 \text{ mm subtends an angle} &= 360/1118° \\
45 \text{ mm subtends an angle} &= 360/1118 \times 45° \\
&= 14.5° \\
50 \text{ mm subtends an angle} &= 360/1118 \times 50° \\
&= 16.1°
\end{aligned}$$

The inlet valve opens 14.5° before TDC and the exhaust valve closes 16.1° after TDC, giving a valve overlap of 30.6°.

Example 2

A bell crank lever has arms 248 mm and 330 mm long (Figure 1.10). The distance between the ends of the arms is 415 mm. Are the arms at right angles? If not, have they opened out or closed together?

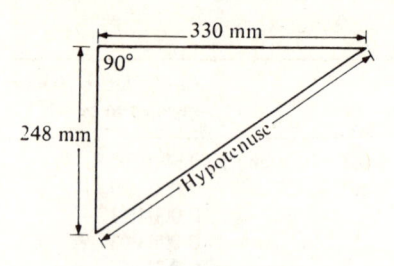

Figure 1.10

Using Pythagoras' theorem for a right-angled triangle:

$$\begin{aligned}
\text{hypotenuse}^2 &= \text{sum of the squares} \\
&\quad \text{on the other two sides} \\
&= 248^2 + 330^2 \\
&= 61\,504 + 108\,900 \\
&= 170\,404 \\
\text{hypotenuse} &= \sqrt{170\,404}\ \text{mm} \\
&= 412.8\ \text{mm}
\end{aligned}$$

The measured distance between the ends of the arms is 415 mm, therefore the angle between the arms must be greater than one right angle.

Example 3

A chassis has a 3.960 m side member, and at right angles to it a 1.120 m cross member (Figure 1.11). Calculate the length of the diagonal and the angle between this and the side member.

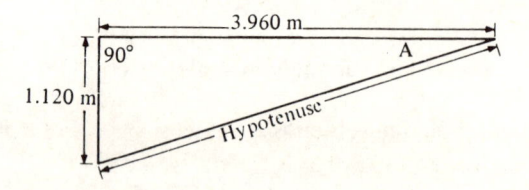

Figure 1.11

Using Pythagoras' theorem for a right-angle triangle:

$$\begin{aligned}
\text{hypotenuse}^2 &= \text{sum of the squares on the other two sides} \\
&= 3.960^2 + 1.120^2 \\
&= 15.6816 + 1.2544 \\
&= 16.936
\end{aligned}$$

Calculations and science

$$\text{hypotenuse or diagonal} = \sqrt{16.936} \text{ m}$$
$$= 4.115 \text{ m}$$

$$\tan \text{ of required angle A} = \frac{\text{opposite side}}{\text{adjacent side}}$$

$$= \frac{\text{cross member}}{\text{side member}}$$

$$= \frac{1.120}{3.960}$$

$$= 0.2828$$
$$= \tan 15° \, 47'$$

Alternatively, using the angle A to calculate the diagonal:

$$\cos \text{ angle A} = \frac{\text{adjacent side}}{\text{hypotenuse}}$$

$$\text{diagonal} = \frac{\text{adjacent side}}{\cos \text{ angle A}}$$

$$= \frac{3.960}{\cos 15°47'}$$

$$= \frac{3.960}{0.9623}$$

$$= 4.115 \text{ m}$$

The length of the diagonal is 4.115 m and the angle between the diagonal and the side member is 15°47′.

Example 4

The parallel portions of a shaft, of diameters 50.14 mm and 60.00 mm, are joined by a tapered surface 70.50 mm long, measured along the tapered surface (Figure 1.12). Calculate the angle between the axis of the shaft and the tapered surface to the nearest degree.

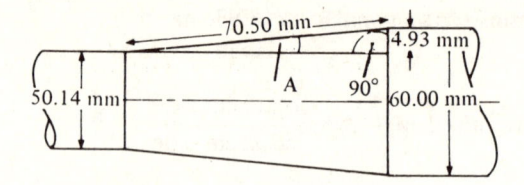

Figure 1.12

Considering the right-angled triangle formed by a line parallel to the axis, a line normal to the axis and the surface of the taper:

$$\sin \text{ of the required angle A} = \frac{\text{opposite side}}{\text{hypotenuse}}$$

$$= \frac{4.93}{70.50}$$

$$= 0.069\,93$$
$$= \sin 4°$$

The angle between the shaft axis and the tapered surface is 4°.

Example 5

The lifting gear on a vehicle body is situated 2.890 m from the pivot (Figure 1.13). If the body contains two loads of 15 kN and 7.5 kN whose centres of gravity are 0.835 m and 1.932 m respectively from the pivot, calculate the vertical force required from the tipping gear to raise the body.

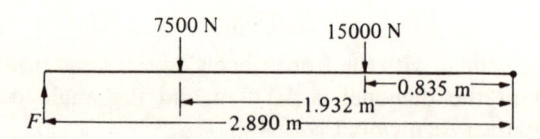

Figure 1.13

Neglecting the weight of the body, which is not given, and taking moments about the pivot:

clockwise moment of tipping force
= sum of anticlockwise moments of loads

Let the required tipping force be F newtons. Then

$$F \times 2.89 = 7500 \times 1.932 + 15\,000 \times 0.835$$
$$= 14\,490 + 12\,525$$
$$= 27\,015 \text{ Nm}$$

$$F = \frac{27\,015}{2.89} \text{ N}$$

$$= 9347.75 \text{ N}$$
$$= 9.348 \text{ kN}$$

The required vertical tipping force is 9.348 kN.

Example 6

A loaded vehicle of wheelbase 4.774 m (Figure 1.14) is weighed by placing first the front wheels and then the rear wheels on a weighbridge, the weights recorded being 9.236 kN and 12.792 kN respectively. Calculate the position of the c. of g. in front of the rear axle.

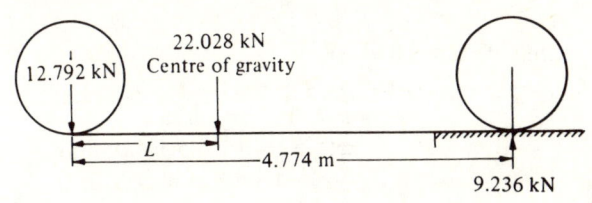

Figure 1.14

Let L be the distance of the c. of g. in front of the rear axle. Then the force acting at L will be the total weight of the vehicle, which is the sum of the weights on front and rear wheels:

$$\text{force at c. of g.} = 9.236 \text{ kN} + 12.792 \text{ kN}$$
$$= 22.028 \text{ kN}$$

Consider the vehicle with the front wheels on the weighbridge. Taking moments about the rear axle:

$$\text{clockwise moment} = \text{anticlockwise moment}$$
$$22.028 \times L = 9.236 \times 4.774$$
$$= 44.093$$

$$L = \frac{44.093}{22.029} \text{ m}$$

$$= 2.002 \text{ m}$$

The c. of g. is 2.002 m in front of the rear axle.

Example 7

(a) *An axle shaft 30 mm in diameter is subject to a shear load of 15.75 kN. Calculate the shear stress.*

$$\text{area of axle shaft} = \pi r^2$$
$$= \pi \times 0.015 \times 0.015 \text{ m}^2$$

$$\text{shear stress} = \frac{\text{shear force}}{\text{cross-sectional area}}$$

$$= \frac{15\,750}{\pi \times 0.015 \times 0.015} \text{ N/m}^2$$

$$= 22\,281\,692 \text{ N/m}^2$$
$$= 22.28 \text{ MN/m}^2 \text{ or } 22.28 \text{ MPa}$$

(b) *A push rod of hollow construction is 11 mm outside diameter and 1.5 mm in thickness. Calculate the compressive stress under a load of 550 N.*

$$\text{area of material} = \pi r_2^2 - \pi r_1^2$$
$$= \pi(r_2^2 - r_1^2)$$
$$= \pi(r_2 - r_1)(r_2 + r_1)$$
$$= \pi(5.5 - 4.0)(5.5 + 4.0) \text{ mm}^2$$
$$= \pi \times 1.5 \times 9.5 \text{ mm}^2$$

where r_1 = inside radius, r_2 = outside radius.

$$\text{compressive stress} = \frac{\text{compressive load}}{\text{cross-sectional area}}$$

$$= \frac{550}{\pi \times 1.5 \times 9.5} \text{ N/mm}^2$$

$$= 12.29 \text{ N/mm}^2 \text{ or } 12.29 \text{ MN/m}^2 \text{ or } 12.29 \text{ MPa}$$

Example 8

An iron block having a mass of 52 kg is cooled from 94°C by 8.5 litres of water at an initial temperature of 21°C. Calculate the rise in temperature of the water and the fall in the temperature of the block during the cooling process.

The specific heat capacity (SHC) of water is 4190 J/kg°C and of iron is 500 J/kg°C. Assuming no heat escapes during the exchange of heat:

quantity of heat lost by the iron block
$\qquad$ = quantity of heat gained by the water

That is,

mass of block × SHC iron × fall in temperature
$\qquad$ = mass of water × SHC water × rise temperature

Let the final temperature of the block and of the water be t°C. Then

$\qquad$ fall in temperature of block = $(94 - t)$°C

$\qquad$ rise in temperature of water = $(t - 21)$°C

Therefore

$\qquad$ $52 \times 500 \times (94 - t) = 8.5 \times 4190 \times (t - 21)$
$\qquad$ $2\,444\,000 - 26\,000t = 35\,615t - 747\,915$
$\qquad\qquad\qquad$ $61\,615t = 3\,191\,915$
$\qquad\qquad\qquad\qquad$ $t = 51.8$°C
rise in temperature of water = $51.8 - 21$
$\qquad\qquad\qquad\qquad$ = 30.8°C
fall in temperature of block = $94 - 51.8$
$\qquad\qquad\qquad\qquad$ = 42.2°C

Example 9

A cylinder has a compression ratio of 8.4 to 1. At the commencement of the compression stroke it is filled with gas at a pressure of 96 kPa absolute and a temperature of 35°C. What should be the pressure of gas, by gauge, at the end of the compression stroke if the temperature rises to 110°C? (Take atmospheric pressure = 101 kPa.)

$$\frac{p_1 V_1}{T_1} = \frac{p_2 V_2}{T_2}$$

where

p_1 = initial pressure absolute = 96 kPa absolute
V_1 = initial volume = 8.4
T_1 = initial temperature absolute = $35 + 273 = 308$ K
p_2 = final pressure absolute (required)
V_2 = final volume = 1
T_2 = final temperature absolute = $110 + 273 = 383$ K

Hence
$$p_2 = \frac{p_1 V_1}{T_1} \frac{T_2}{V_2}$$

$$= \frac{96 \times 8.4 \times 383}{308 \times 1}$$

$$= 1002.8 \text{ kPa absolute}$$

gauge pressure $= 1002.8 - 101$
$= 901.8$ kPa

Chapter 2
Drawing

2.1 Equipment

For satisfactory engineering drawing the following equipment is necessary: drawing-board, paper, pins, tee-square, 45° and 60° transparent set-squares, 30 cm rule graduated in 1 mm, 2H, H and HB pencils, sandpaper block, rubber, bow compasses and 15 cm compasses – both with shouldered needle points, and the latter with a substantial joint. Dividers and ink bows are not required for this work.

2.2 First-angle projection

The aim of an engineering drawing is to indicate as clearly and simply as possible all the details and dimensions of an object. For this purpose three views are usually required – the front elevation, the plan and the end elevation. In the first-angle method the views are drawn on the opposite side from which they are projected. The plan, viewed from above, is projected and drawn below the front elevation; the end elevation is drawn on the opposite side of the front elevation from which it is viewed.

2.3 Third-angle projection

In the third-angle method the views are drawn on the same side from which they are seen. The plan viewed from above is drawn above the front elevation; the end elevation is drawn on the same side of the front elevation from which it is viewed.

2.4 Sectional views

Sectional views are sometimes necessary to show hidden internal details. It should be imagined that the object has been sawn through on the line of section and the portion nearer the direction of view removed. The 'sawn' portions are then represented by evenly spaced 45° section lines, whilst the 'uncut' parts are shown by outline only.

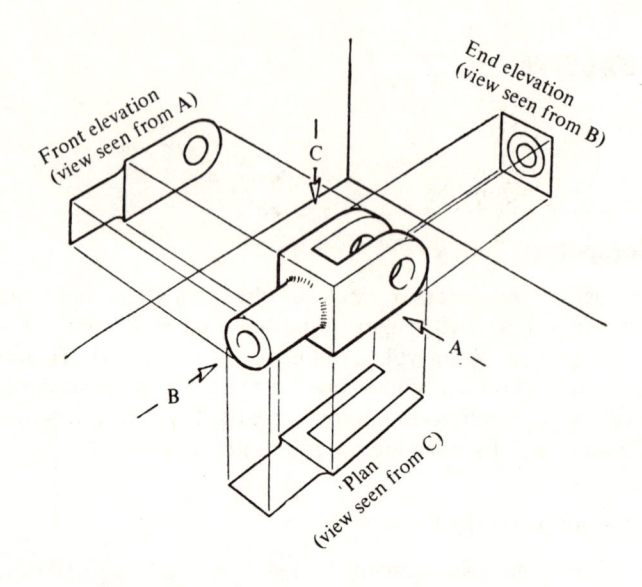

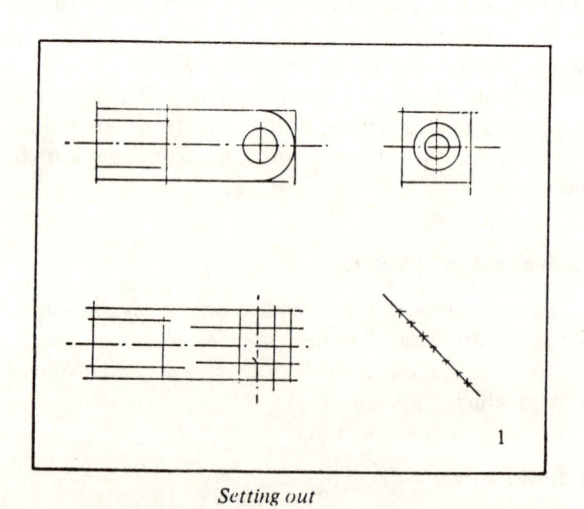

Figure 2.1 First-angle orthographic projection

Drawing

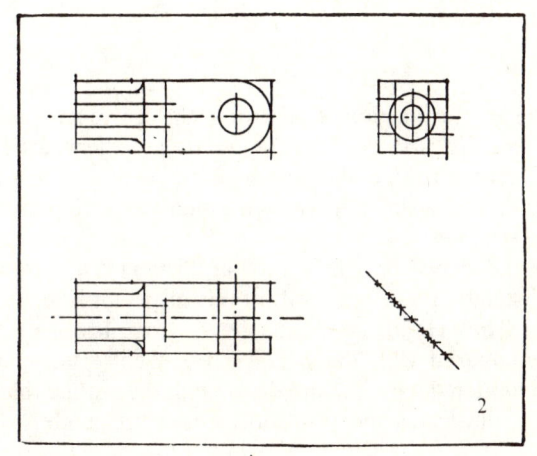

Filling in

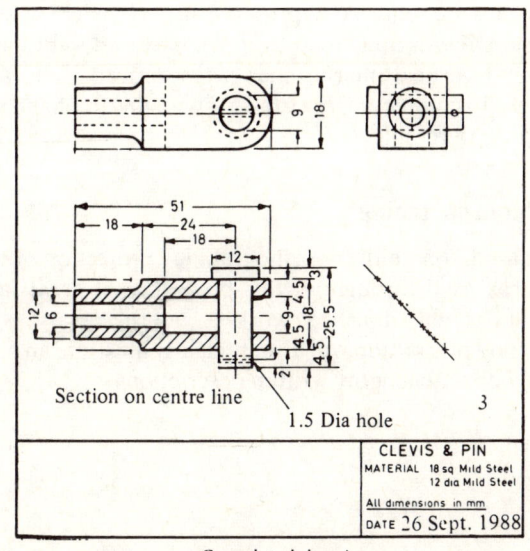

Section on centre line
1.5 Dia hole

CLEVIS & PIN	
MATERIAL	18 sq Mild Steel
	12 dia Mild Steel
All dimensions in mm	
DATE	26 Sept. 1988

Completed drawing

29

As there is no hidden detail to reveal in shafts, keys, webs, gear teeth, balls, rollers, bolts, nuts, screws and pins, these objects are not section lined when 'cut' longitudinally.

2.5 Method

After setting out the position of each view, the centre lines and all circles and main arcs are first drawn. The views are then carried forward together. The faint projection lines are actually drawn only in the area of each view, and a 45° line is used when projecting from the end elevation to the plan.

When all the views are lightly filled in, distinct firm outlines of even thickness throughout are required; in this respect it is an advantage to use a softer lead (H) in the compasses – where less pressure can be applied – compared with the pencil (2H). Centre lines, dimension, section and hidden detail lines are drawn half the outline thickness.

Figure 2.1 illustrates the procedure for a first-angle orthographic projection.

2.6 Printing and dimensions

Faint guide lines are required to preserve an even height when printing. The dimension figures must read from the base and right-hand side for horizontal and vertical measurements respectively. The arrowheads must be small. In completing the drawing with this freehand work, an H or HB pencil is preferable.

2.7 Freehand sketching

Sketches based on either orthographic projection or pictorial projections, as used throughout this book, are of great importance. Elaborate and highly finished work is not necessary; simple line sketches in good proportion will usually clearly illustrate an operation or assembly and obviate lengthy written descriptions.

Chapter 3
The vehicle structure

3.1 The chassis frame

In this construction, mainly employed for commercial vehicles, the chassis frame must maintain the working assemblies in their correct position and provide a mounting for the body, which adds little, if anything, to the strength of the structure.

The usual commercial vehicle construction consists of two heavy-gauge channel-section side members connected by channel- or box-section cross members riveted, bolted or welded in place. The side members are usually straight and may have an increased depth of section towards the centre where the greatest bending loads occur.

The principle is suited to low-volume production, and a variety of bodies for specialized duties can be accommodated on the same chassis.

3.2 Integral construction

On most cars the separate chassis frame and body are replaced by a unitary welded construction of mild-steel pressings, which may, in some cases, incorporate a rudimentary 'chassis' in the form of channel-, box- or top-hat-section longitudinal and transverse members. Where large-scale production justifies the necessary equipment, this mono-construction offers a rigid unit with considerable saving in weight and cost (Figure 3.1).

Intermediate designs are sometimes employed, where a light-section chassis frame relies to a considerable extent upon a separate pressed-steel body for its strength and rigidity. Alternatively, the chassis 'frame' may consist of a sheet-steel platform, braced by channel-section members, which forms the floor of the car body.

With the thin gauge material used in integral construction, the design must avoid concentrated loads and ensure that the stresses are spread over a large area. Protective treatment against corrosion is essential for acceptable longevity. The front and rear of the construction may be arranged to progressively crumple under a severe impact and reduce the

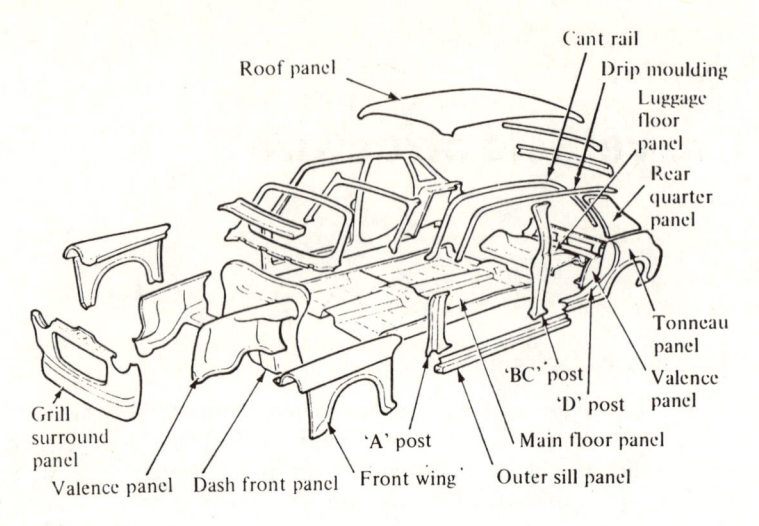

Figure 3.1   Body panels

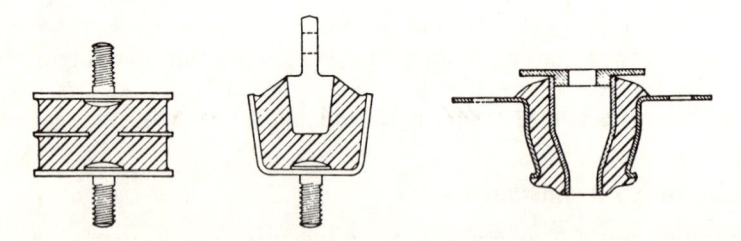

Figure 3.2   Flexible mountings

deceleration or acceleration imposed on the belted passengers in the (hopefully) intact central area. Aerodynamic analysis is required for the body profiles to provide performance, economy and stability at high speed.

The suppression of noise, vibration and harshness is an important department of body design. Rubber-mounted subframes are widely used for independent front suspension and often include the engine-gearbox or engine-transaxle unit. Similar rear subframes may carry the rear independent suspension and include the final drive unit. Sound-damping materials – rubber, bituminized paper, resin-bonded fibre, foam fillings etc. – are all widely applied.

The engine unit is normally carried on three flexible mountings, which must not only insulate vibration but also support a torque

The vehicle structure

reaction equal and opposite to the driving torque. This controlled engine movement must be accommodated by the exhaust, water and fuel connections and the clutch, accelerator and other linkages. The electrical insulation of the rubber mountings necessitates an engine–body 'earthing' lead (Figure 3.2).

3.3 Accident repair

Where distortion is suspected first check the steering geometry, the wheelbase measurements and the wheel alignment front to rear.

With the wheels removed, support the vehicle (or chassis) above a level floor so that two corresponding datum points – nearside and offside – at the rear have equal height. If the heights of a front pair of datum points then cannot be equalized, the chassis is twisted. With the vehicle supported in this way, at the maker's datum height, vertical alignment can be checked against the maker's heights for specified points.

Check for squareness by projecting datum points on to the floor with an accurately balanced plumb-bob and fine cord, the vehicle being

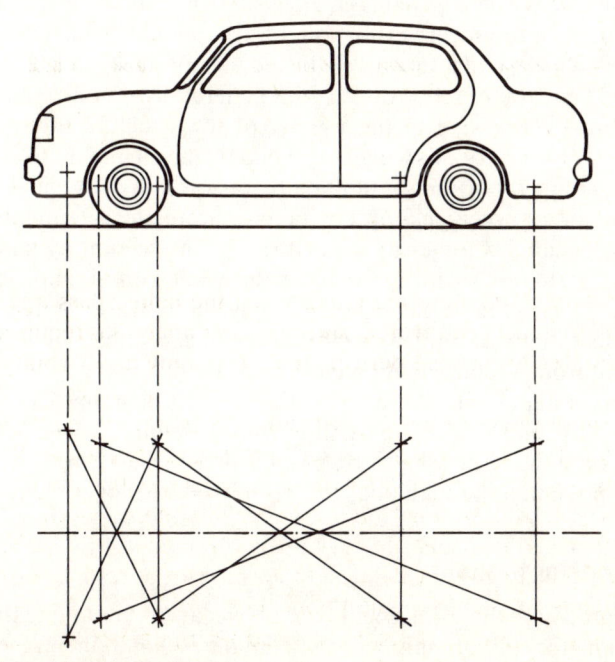

Figure 3.3   Alignment check

supported as before and any necessary assemblies removed for access. Diagonal lines between corresponding pairs of datum points should be equal and all should intersect on the same centre line. As a further aid in locating the distortion, join corresponding points; the midpoints of these lines – crossing the centre line at 90° – should also coincide with the centre line (Figure 3.3).

Measurements can also be checked against the manufacturer's horizontal dimensions for specified lengths.

Alignment of subframes may be checked against the maker's distances from specified datum lines or faces.

The use of a body jig, with the maker's attachments for a particular damaged vehicle, greatly facilitates accident repair work. The brackets locate on the major datum points on the body, securing it to the jig for whatever straightening operations are required and also immediately indicate distortion – without the necessity for the alignment checks otherwise required.

Repair

When the bending is not too great, a chassis frame can be straightened by using a heavy beam, chains and jacks. Slight bending can be rectified cold, but for more severe distortion the area should be heated to cherry red (750–850°C) and afterwards allowed to cool slowly in air.

The correction of a body shell with general distortion is not practicable. However, owing to the rigidity of the structure the damage is frequently localized; this will be revealed by the alignment tests.

The damaged portion should first be restored to its original position with the aid of body jacks and by taking careful measurements. It can then be decided whether to repair the existing section by panel-work techniques, to replace the entire panel or – what is often simplest – to cut out and replace the damaged portion only. In this case a patch, formed from sheet metal of similar specification and gauge, or a corresponding piece cut from a replacement panel can be welded in.

Where a panel must be replaced the maker's diagrams will facilitate tracing the joints. Spot welds should be carefully drilled out, penetrating only the upper sheet. Butt welds must be cut through; in many cases paint and body solder has to be removed from over the joint.

3.4 Engine position

Compared with the front-engine rear-drive layout, mounting the engine adjacent to the driving wheels eliminates components, improves traction and permits a less obstructed floor. It necessitates independent suspension of the driven wheels – except where, occasionally, the De Dion dead rear axle is used.

The front-engine front-wheel-drive (FWD) system offers the optimum body and luggage space and good handling characteristics, but has the complication of driven steered wheels. In addition, weight transfer effects may restrict effective rear braking. Driven front wheels deliver the tractive effort along the steered path, and on adverse terrain this assists in surmounting obstacles. When the vehicle is cornering, a centripetal vector of the tractive force allows a slightly greater speed – or safety margin – than with rear-wheel drive. Steering and suspension geometry, however, must guard against any instability on throttle lift-off or braking, when cornering near the limit of adhesion.

In the rear-engine rear-drive layout the provision of stable handling characteristics is more difficult. In addition, the front wheel arches obstruct passenger space and the front luggage compartment may not be so convenient; controls and interior heating are more complicated. This layout is virtually obsolete on cars, but it has advantages for public service vehicles where it allows a low floor level throughout the passenger accommodation.

Chapter 4
Suspension

Road shocks are absorbed by the suspension springs, for as the wheels rise and fall with the road undulations the spring forces are increased or decreased by a comparatively small amount and the resulting movement of the body is consequently slight.

The flexibility or *rate* of a spring measured in kN/m or N/mm is the force required to produce a deflection of 1 m or 1 mm. Simple springs, whether of the leaf, helical or torsion-bar type, have a constant rate: they follow *Hooke's law*, which states that the deflection is directly proportional to the force producing it.

For maximum comfort – 'soft suspension' – a low-rate spring is required. This necessitates a large static deflection, large spring movements and a long period of natural (undamped) vibration. The aim of car design is towards a natural spring frequency of 1 Hz (Figure 4.1).

The lowest possible unsprung weight is desirable in the suspension system to reduce the inertia forces of the wheels and axles. The springs can then more readily maintain the wheels in contact with the road surface – for maximum control – and less force is transmitted to the chassis. Independent wheel suspension offers important advantages in permitting large and accurately controlled wheel movements and reduced unsprung weight.

The three basic suspension movements of the body are *bounce* or *float*, a vertical movement of the complete body; *pitch*, a rocking-chair movement about a transverse axis; and *roll*, a movement about a longitudinal axis produced by centrifugal force when the vehicle is cornering.

4.1 Leaf springs

The semi-elliptic type of leaf spring, though virtually 'flat' in use, is employed principally on commercial chassis. It serves as a spring and also for axle guidance, but this becomes imprecise with a low-rate spring.

Suspension

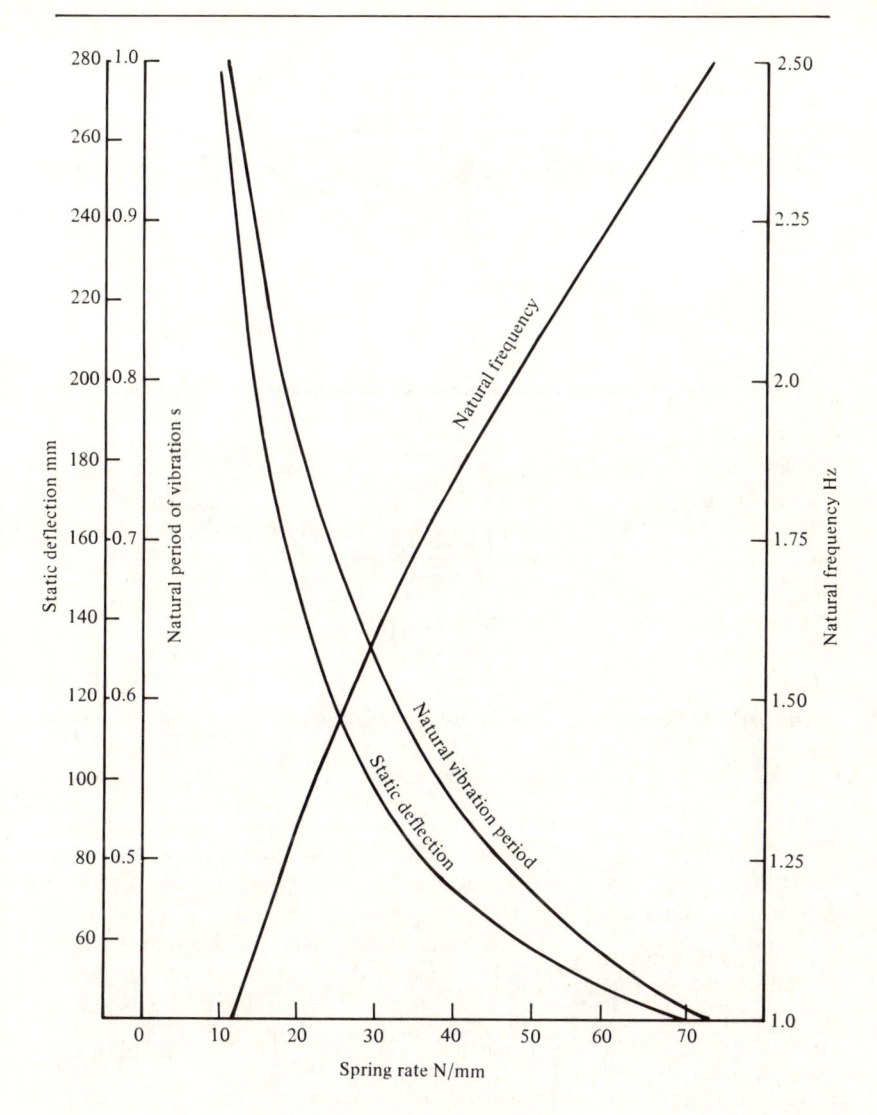

Figure 4.1   Spring characteristics for 3 kN load

In the laminated spring (Figure 4.2), the leaves are graduated in length from the main leaf to provide a uniform stress throughout the material. The leaves are located by a centre bolt or a series of interlocking dimples and are secured to the axle by U-bolts. Spring clips clamp the leaves together and prevent overloading of the master leaf

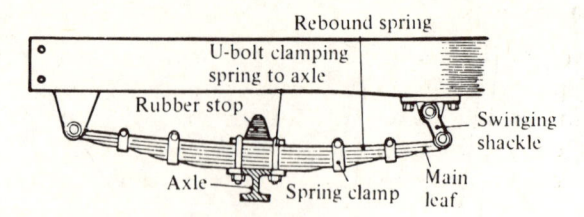

Figure 4.2 Semi-elliptic leaf spring

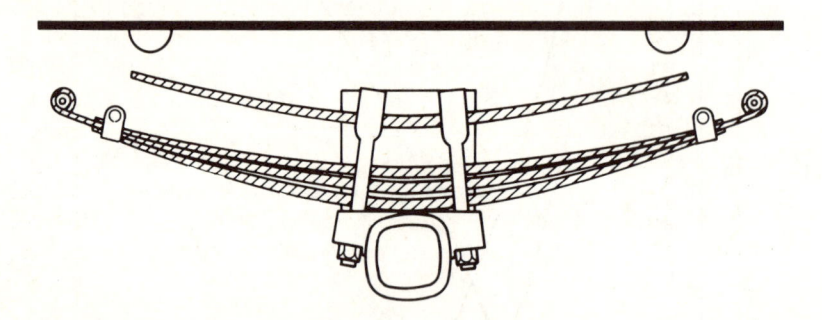

Figure 4.3 Tapered-leaf main and helper springs

during rebound, when otherwise the leaves would separate and the whole spring force would be provided by the main leaf. A swinging shackle, or a slipper device, is necessary to accommodate the difference in centre distance between the spring eyes during deflection. The eyes can be fitted with rubber, self-lubricating plastic or phosphor-bronze bushes. Heavy-duty metal bushes may be threaded – on threaded pins – to increase the bearing area. The interleaf friction dampens vibrations but reduces flexibility and is unpredictably variable. When leaf springs are used for car rear suspension it is usual to minimize interleaf friction by plastic interleaves or plastic buttons and to sheath the axle mounting in rubber.

A tapered leaf can be lighter than laminations because of more even stress distribution. There may be a single leaf, or a main leaf with one or two subsidiary leaves, bearing only at the centre and ends to minimize friction. The system is popular on some commercial chassis (Figure 4.3).

It should be noted that where semi-elliptic longitudinal springs are employed, some twisting of the springs must occur when one end of the axle is deflected.

Suspension

4.2 Coil springs

For a given energy storage coil springs require only about half the mass of a leaf spring. In addition, a lower proportion of the mass is unsprung. Coil-spring action is mainly torsional with some bending. Torsion bars reduce the mass and unsprung proportion even further but necessitate special end fixings with height adjustment.

4.3 Variable-rate springs

A rising-rate spring allows the benefit of soft suspension for small movements whilst keeping the total deflection within acceptable limits. Helper leaf springs, contacting chassis pads only after some deflection of the main leaf springs, are commonly used on commercial vehicles (Figure 4.3). Arranging the end of the main leaf to slide in a slipper pad, instead of a swinging shackle, also increases the rate as its effective length decreases. As coil springs are normally of the same gauge and form throughout, a rising rate can only be achieved by the geometry of the suspension lever or by the helper principle. A rising rate is inherent with a simple air strut, but the static deflection would alter with the temperature of the contained air.

4.4 Rubber springs

Rubber bushes and mountings act to some extent as springs and are widely used to isolate vibration and road noise. For suspension systems these flexible connections must provide compliance for any divergence of movement arcs without permitting inaccurate geometry.

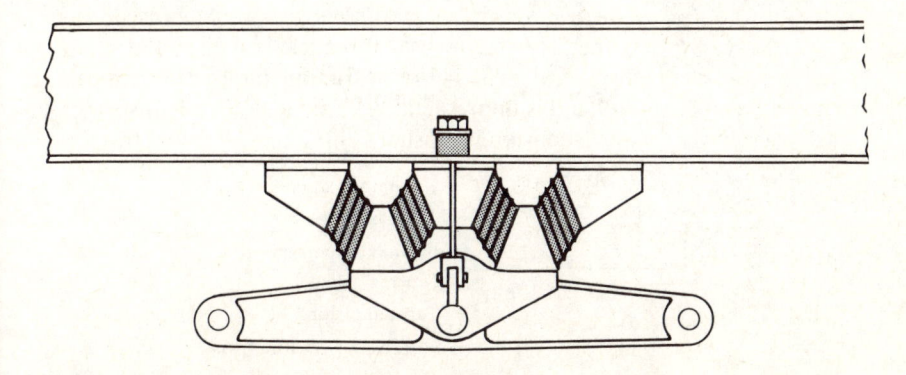

Figure 4.4 Commercial rubber suspension with bonded-in steel plates

Rubber springs are used for both light vehicles and heavy commercial applications. The mountings are arranged to ensure a wedging action under deflection; this progressively changes shear loading of the material into the much stiffer compression loading, thus providing a rising rate (Figure 4.4).

4.5 Interconnected suspension

When the front and rear suspension each side of the vehicle are connected so that an increase in the loading of one is transmitted to the other, the effective spring rate is lowered and body pitching is reduced.

Coil springs with mechanical linkage have been used. More elaborate systems employ rubber cone springs (Figure 4.5), or compressed nitrogen springs with flexible diaphragms and hoses communicating the pressure from front to rear via a water-antifreeze mixture (Figure 4.6).

4.6 Self-levelling suspension

A self-levelling function is valuable because it makes the full suspension movement available at all times, irrespective of loading. The compressed air system is attractive for coaches, where passenger comfort is a

Flexible hose to longitudinal connected unit

Antifreeze fluid (alcohol-water-anticorrosive solution) pressurized to 775-800 kN/m^2

Rubber in compression and shear

Fluid chamber

Impact valve

Damper flap-valve assembly (half-views)

Rebound valve

Low-speed damping bleed hole

Separator plate

Fluid chamber

Flexible diaphragm

Tapered cylinder

Tapered piston

Strut from suspension arm

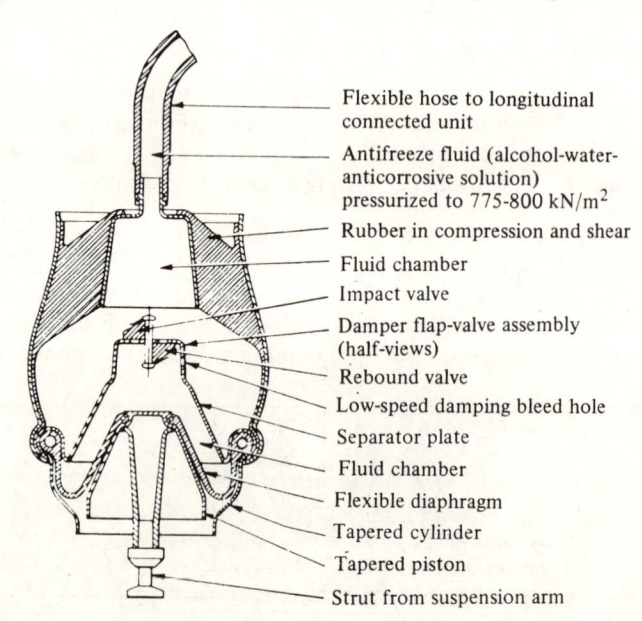

Figure 4.5   Interconnected system: rubber suspension unit

Suspension

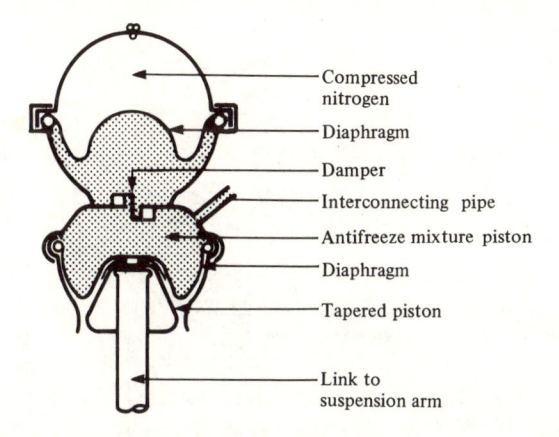

Figure 4.6 Interconnected system: gas suspension unit

priority and an air system is already employed for power braking. Air is pumped into or released from air springs to maintain a constant static deflection. The springs consist of chambers and pistons sealed by flexible folding diaphragms. Their shapes are contoured to provide increasing piston area and the desired spring rate increase with deflection.

Cars use a hydro-pneumatic layout where the spring is compressed nitrogen contained in a chamber and diaphragm, again contoured to give the desired rising rate. Height control is obtained by adding or releasing high-pressure oil between the diaphragm and a piston moving with the suspension. The hydraulic pump can power the steering, brakes, gear change or clutch and the system can provide additional body height for snow, adverse terrain or wheel changing (Figure 4.7).

All self-levelling valves must incorporate a designed time delay so that they do not respond to transient alterations of the static height or the normal rapid bump and rebound of the individual suspension units.

4.7 Active suspension system

The active suspension system can be considered as a further development of both the interconnected and self-levelling systems. Using a similar hydro-pneumatic layout, very rapid additions and subtractions of the hydraulic fluid between suspension piston and gas diaphragm are made. Sensors monitoring body height, roll, pitch, brake dive and acceleration squat act with minimum delay to operate the valving and maintain a stabilized body plane. Again the valving arrangements must

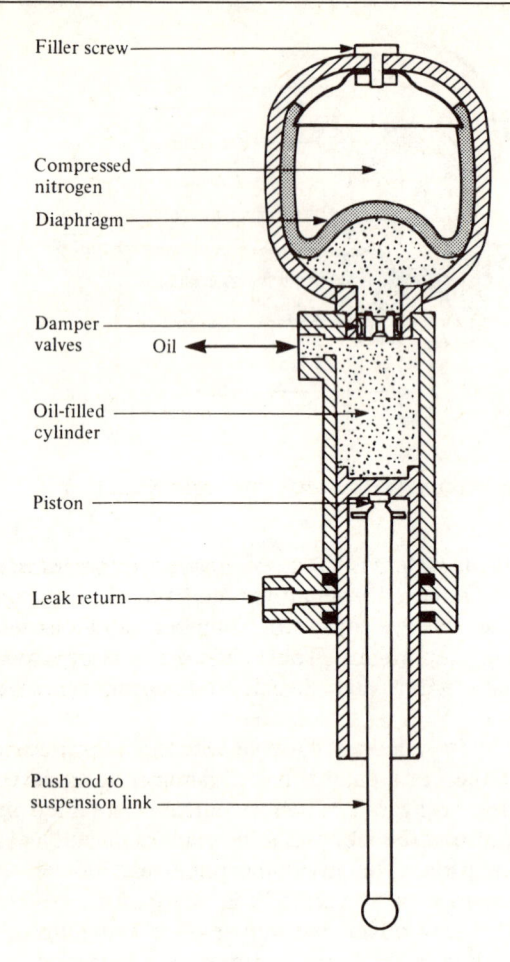

Figure 4.7   Self-levelling suspension unit

ensure that the hydraulic system does not respond to the bounce and rebound of an individual wheel. This is an area where electronic control systems are likely to prove more comprehensive and economical.

4.8   Viscous damping

After the initial deflection of any undamped spring, oscillations of the chassis on the suspension are set up and a series of road bumps can build up a dangerous resonant bouncing of the sprung mass. These oscillations can be *damped* by converting their energy into heat, either by the

Suspension

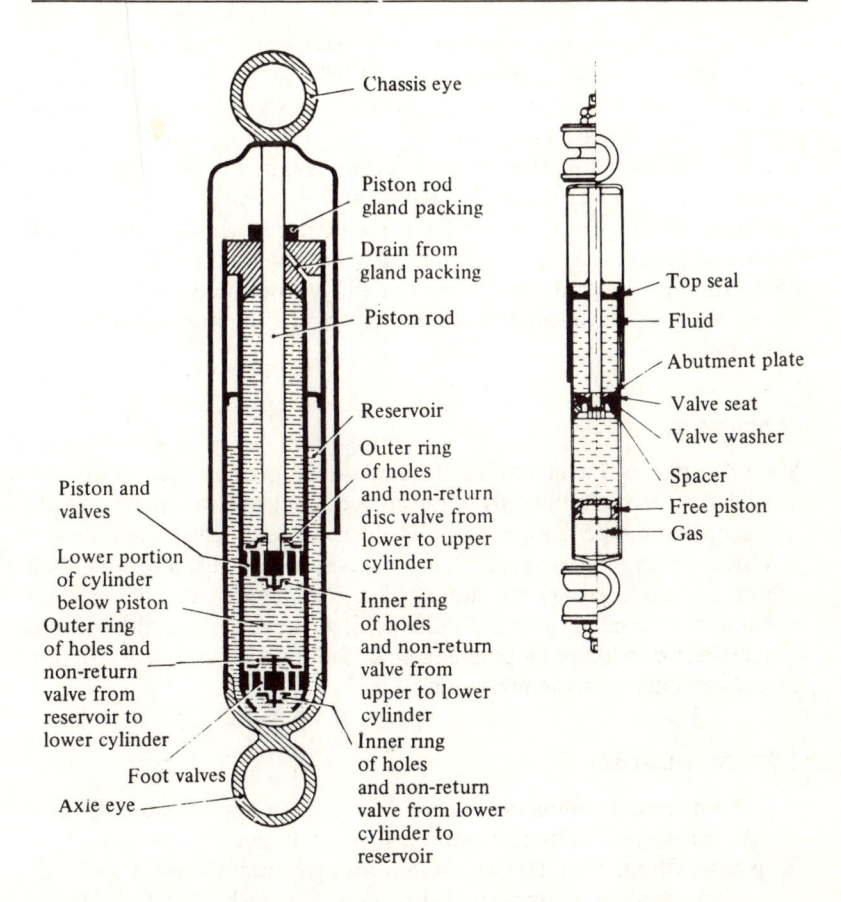

Figure 4.8   Hydraulic damper

'solid' friction of the older friction damper, or by the fluid resistance of the hydraulic damper. In addition, the damper restricts any rapid or excessive vibrations of the unsprung mass.

The direct-acting telescopic damper, or the parallel- or opposed-piston types of hydraulic damper, all employ a piston or pistons to displace oil from one chamber to another through spring-loaded one-way valves and calibrated orifices (Figure 4.8). The fluid resistance is proportional to the square of the speed of flow and so increases rapidly with the speed of the suspension movement. A differential action is provided by the valves so that the bump resistance is low – allowing the suspension to rise easily and absorb small shocks – but the rebound

43

resistance is much greater and so checks oscillations. Arrangements of bleed holes and valve spring loadings enable the required damper characteristics to be obtained. In the telescopic damper various methods are utilized to accommodate the differing volumes of fluid above and below the piston. This is a consideration in the MacPherson strut where a large-diameter piston rod is needed to accept the guidance loads. When monotube construction is used – more at risk from a dented casing – gas at some 2.5 MPa under a floating piston can be used. One alternative is the use of a finely emulsified oil-gas mixture where the volume variations are absorbed by the compression of the gas bubbles.

Maintenance

Most dampers are sealed units. If topping-up is possible, care is needed to exclude dirt and to fill only to the correct level with the specified fluid whilst operating the damper to expel air from the pressure chambers.

When testing by hand the bump resistance is usually less than rebound, and air may be indicated by sponginess or the ease of movement. Hand testing is inadequate, however, since the viscous resistance is velocity dependent and the piston and seal friction complicates comparison with a new unit.

4.9   Anti-roll bar

A U-shaped bar, with ends linked to the axle or wheel assembles and usually turning in rubber mountings on each side of the chassis, is frequently fitted. With the suspension acting so that the body and axle remain parallel the bar turns in its brackets, but any body roll is resisted by the torsion set up in this anti-roll bar. The front anti-roll bar often doubles as a stay for the suspension arms on strut suspension. The penalty for using an anti-roll bar is the increased rate for single-wheel bump and rebound.

4.10   Suspension geometry

When the vehicle is cornering, centrifugal force acts on the sprung mass against the resistance of the suspension springs and roll bars. The amount of roll depends upon the stiffness of the springs, the width of their base and the position of the roll axis – a longitudinal axis joining the front and rear roll centres. The location of a roll centre is determined entirely by the suspension design.

Body roll, by altering the direction of front or rear wheels, can introduce roll steering and deflect the vehicle from course. The presence and degree of roll steer depends upon the suspension layout. Greater roll stiffness is always promoted at the front – often by the roll bar – to provide understeer characteristics.

Braking and accelerating forces also act on the sprung mass and necessitate geometry in the suspension linkage that will oppose brake dive and acceleration squat. Suspension design is always a compromise between various conflicting requirements.

Chapter 5
The front-wheel assembly

5.1 Ackerman linkage

When the vehicle is cornering, true rolling motion without sliding can be obtained only if the produced axes of all the wheels intersect in a single point. Since the planes of rotation of the rear wheels are fixed, this point must lie on the produced centre line of the rear axle.

The front wheel on the inside of the corner has therefore to be turned through a greater angle than the outer wheel, and this is approximately accomplished by what is generally called the Ackerman linkage (Figure 5.1). The track arms are arranged so that their centre lines, when in the central or straight-ahead position, if produced would meet on the centre line of the vehicle near the rear axle. These considerations apply to vehicles cornering at low speed. At increased speed cornering side force causes the centre of turning to move ahead of the produced centre line of the rear axle. Whilst some cars and lorries (including those with twin-steered axles) continue to use the Ackerman layout, most cars employ a reduced-Ackerman or parallel wheel movement to increase the slip angle of the outer wheel. With any type of Ackerman linkage it follows that the track rod will be shorter than the distance between the swivel-pin axes if it lies behind the axle beam, and longer if it lies ahead.

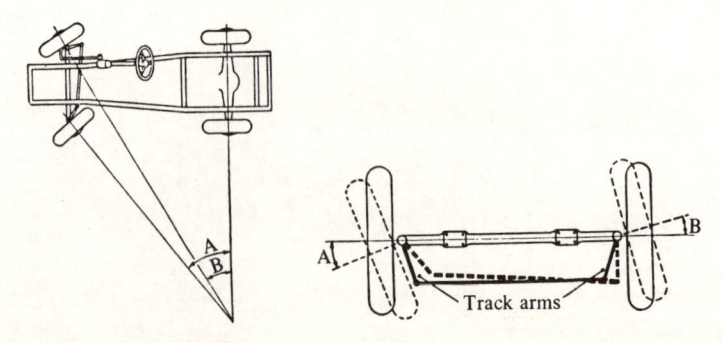

Figure 5.1   Ackerman linkage

Checking the cornering wheel relationship – toe-out on turn – requires the manufacturer's data, since the relative angles vary considerably.

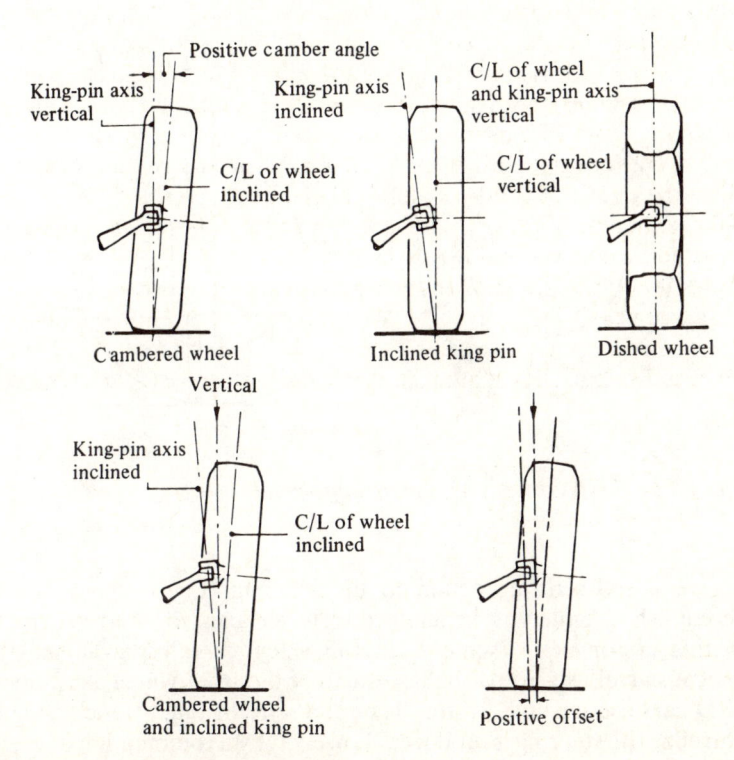

5.2 Centre-point steering

Centre-point steering occurs when the produced axes of the swivel pins intersect the ground at the centre point of the tyre contact area. With perfect centre-point steering, the ground resistances due to uneven surface or braking can produce no turning moment of the wheel about the swivel axis.

Centre-point steering can be obtained by (a) inclined or cambered wheels, (b) inclined swivel axis (king-pin axis), (c) dished wheels or (d) a combination of (a), (b) and (c) (Figure 5.2).

Usually in modern practice the centre of the tyre contact area lies outside the swivel axis to give up to 25 mm of positive offset. When the

Figure 5.2   Centre-point steering

wheels are steered the difference between the turning moments on each stub axle has a self-centring action which is effective in preventing wobble. A number of cars, especially those with FWD or diagonally split braking systems, use a negative offset of 10–25 mm. This minimizes the effect of one brake or tyre failure.

In modern design, and especially with low profile tyres, the wheels are as nearly vertical as possible under running conditions. Typically they have an outward or positive camber of about 1°, though some cars have a slight negative camber which can improve the cornering power of the tyre. The steering-axis inclination (king-pin inclination (KPI)) is usually about 8°.

5.3 Castor angle

The produced swivel axes are generally arranged to meet the road ahead of the centre of tyre contact area, by an amount termed the *trail*, in order

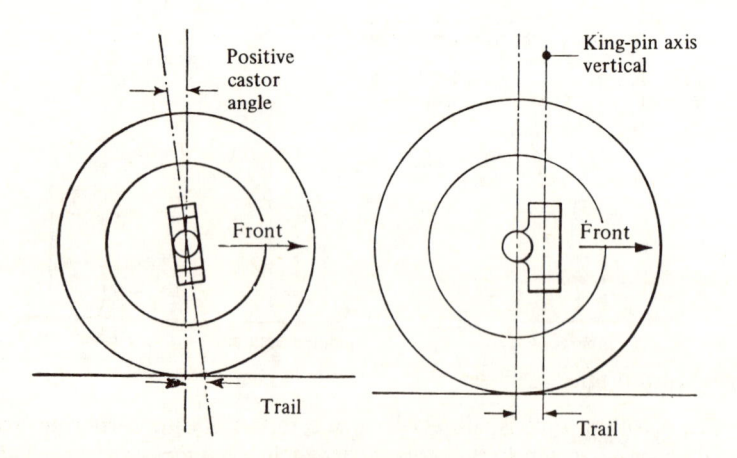

Figure 5.3 Positive castor angle, trailing wheel

to give a self-centring action to the steering (Figure 5.3). This is accomplished either by inclining the swivel axes so as to provide a positive castor angle (some 2°), or in a few cases by arranging the vertical swivel axes to be slightly ahead of the front-wheel axes. Some FWD cars use negative castor. The total self-aligning torque acting to centralize the steering is produced, however, by a combination of wheel offset, KPI, camber, castor and the tyre characteristics.

5.4 Wheel alignment

The front wheels of a rear-driven vehicle usually have a tendency to run out or move so that the track between the rims would be greater at the front than at the rear. To compensate for this and so give parallel running, a toe-in is usually given (Figure 5.4). Front-wheel-drive vehicles have the opposite tendency, and a similar toe-out is customary.

Wheel alignment is equally applicable to the rear wheels when these are not constrained by a live or dead axle. Since alignment is affected by many factors, including the type of suspension and tyres fitted, some makers specify steering geometry reversing normal criteria, e.g. toe-out on some FWD cars.

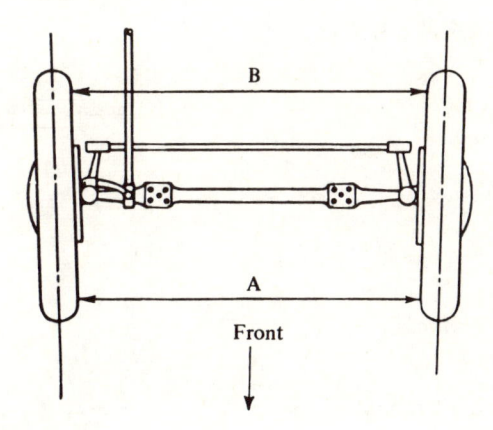

Figure 5.4   Toe-in (B − A)

5.5 Slip angle

When a rolling tyre is subject to a side thrust due to cornering, road camber or side wind, the area of tread in contact with the road is distorted. The actual path of the tyre along the ground makes an angle with the plane of rotation called the slip angle, so that the wheel does not travel in the exact direction of its rotation (Figure 5.5).

For a given tyre and wheel loading the slip angle is proportional to the side force up to about 5°. It reaches a maximum of around 12°, then decreases as the tyre deformation develops into actual slip on the road and sliding occurs. For this reason the slip angle can be more properly referred to as the *drift angle.*

The value of the slip angle has important effects on the steering characteristics of the car. If the slip angle for the rear tyres is greater

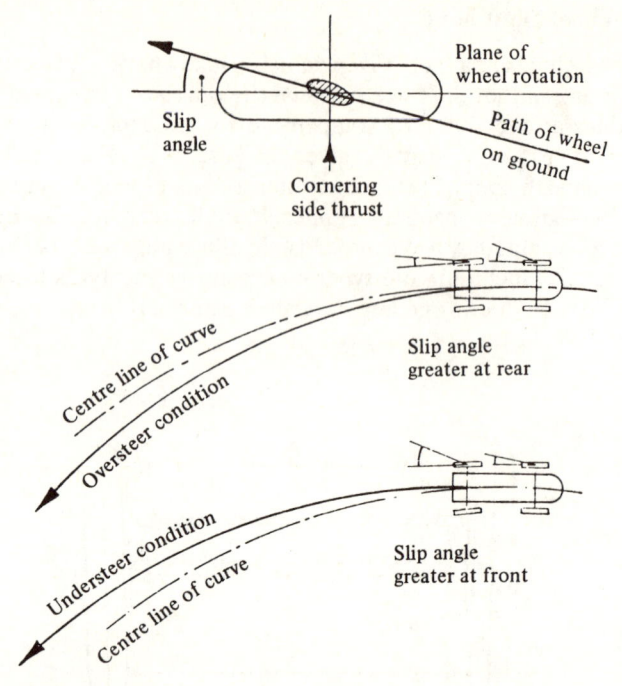

Figure 5.5 Slip angle

than that for the front tyres, the car will oversteer when cornering; that is, it will turn more sharply into the curve, and the steering wheel has to be 'turned back' to correct this tendency. On the other hand, if the slip angle is greater for the front tyres than for the rear, the car will understeer and must be 'held in' to the curve. The latter condition makes for stability since the steering-wheel rotation is in the same direction throughout. Slight understeer is also required for a vehicle to have straight-line stability under the influence of road camber or side wind.

The steering characteristics are determined by the design of the vehicle, but the conditions under which it operates can have an overriding effect. For example, heavy rear loading or low rear-tyre pressures will promote oversteer.

5.6 Cornering power

Cornering power, an important criterion of tyre design determined by the tyre's construction and aspect ratio, is defined as the cornering force

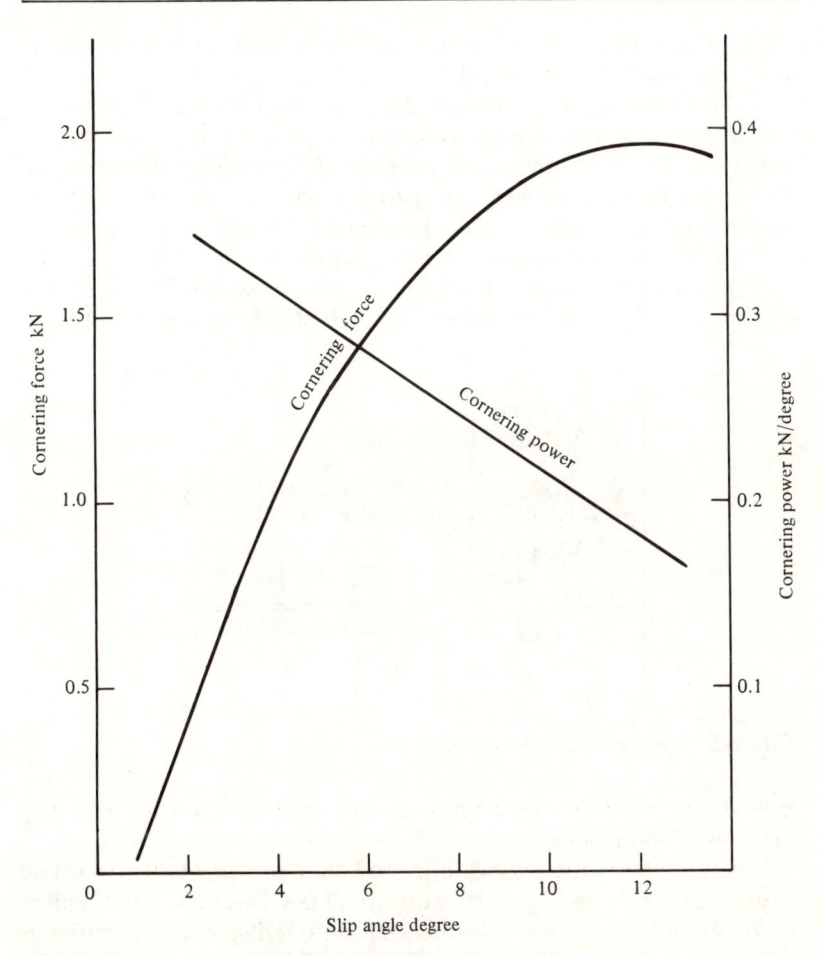

Figure 5.6 Cornering force and power for 3 kN load

divided by the slip angle producing it (Figure 5.6). Cornering power decreases steadily as the slip angle increases and also decreases with increasing positive camber; slight negative camber improves the cornering power.

5.7 Axle-beam construction

The axle beam is a steel forging, usually of I-section between the springs to withstand the vertical bending load, and of oval or circular section in the outer portions better to resist the braking torque. The spring platforms are drilled to take the spring clamp bolts and have a dowel hole to locate the head of the centre bolt.

In the reversed-Elliot construction, usual on commercial vehicles, the forked stub-axle forging turns on a swivel pin of hardened steel, secured in the axle-beam eye by a tapered cotter. A hardened-steel washer or ball-thrust bearing supports the vertical load, and phosphor-bronze bushes form the bearing surfaces for the swivel pin (Figure 5.7).

Each stub axle has a track arm secured to it and one of these, usually on the same side as the steering box, may be extended to form the steering arm. The axle beam may carry stops which engage with the stub

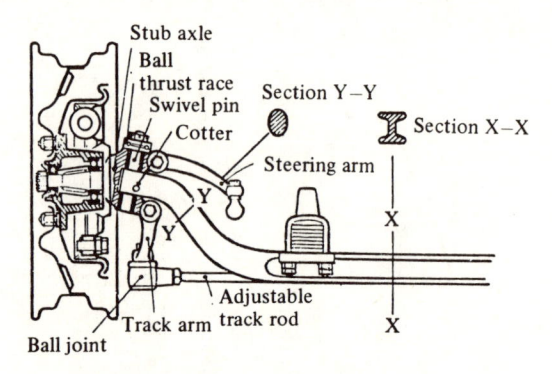

Figure 5.7 Reversed-Elliot construction

axle and restrict the steering movement to prevent the wheels fouling the chassis on full lock.

When semi-elliptic springs are used they are pivoted at the end nearest the steering box so that the arc of movement of the drag link follows more closely the movement of the steering arm as the springs deflect. The braking torque is transmitted through the springs to the chassis.

5.8 Independent front suspension

The leaf-sprung beam axle has long been unacceptable except on commercial chassis. Low-rate springing requires large accurately controlled movement without wheel tilting – producing gyroscopic reaction – or the inertia effect of a large unsprung mass. A wide spring base promotes front roll resistance and understeer characteristics. Better space usage results from an engine position unrestricted by the necessity for axle clearance.

Apart from the use of live front axles on some four-wheel drive

(4WD) vehicles, independent front-wheel suspension (IFS) is mandatory for front-wheel drive. Current types of IFS comprise the double transverse link or wishbone, and the transverse link and strut or MacPherson.

5.9 Double transverse link

The lengths and axes of the links are chosen for the best compromise of steering and suspension geometry (Figure 5.8). A longer lower member maintains more constant track at the cost of camber change – though a slight negative change on the outer wheel improves cornering power.

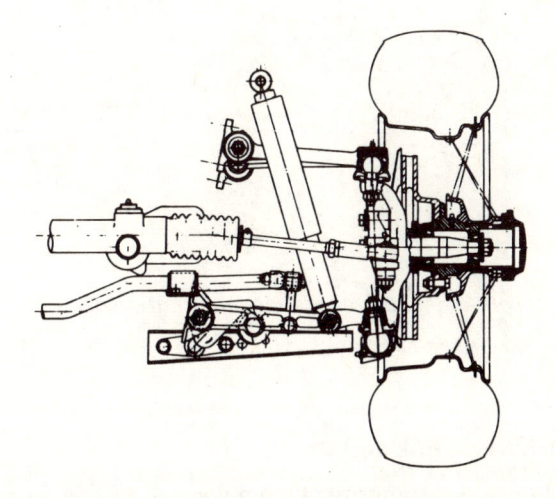

Figure 5.8 Double-transverse-link independent front suspension

Converging the upper and lower axes of the links to the rear controls brake dive at the expense of castor variation.

The links are adequately splayed or stayed to resist braking torque. A coil spring with coaxial damper acts between the body and the upper link for FWD and the lower for rear-driven cars. Alternatively a longitudinal torsion bar can be driven from the suspension link.

Ball-and-socket joints used between the stub axle (or FWD swivel hub) and the suspension arms eliminate the swivel-pin arrangement. On FWD vehicles a constant-velocity joint, which may have a plunge facility to eliminate sliding splines, is located on the axis of the ball joints to provide for steering and suspension movement (Figure 5.9).

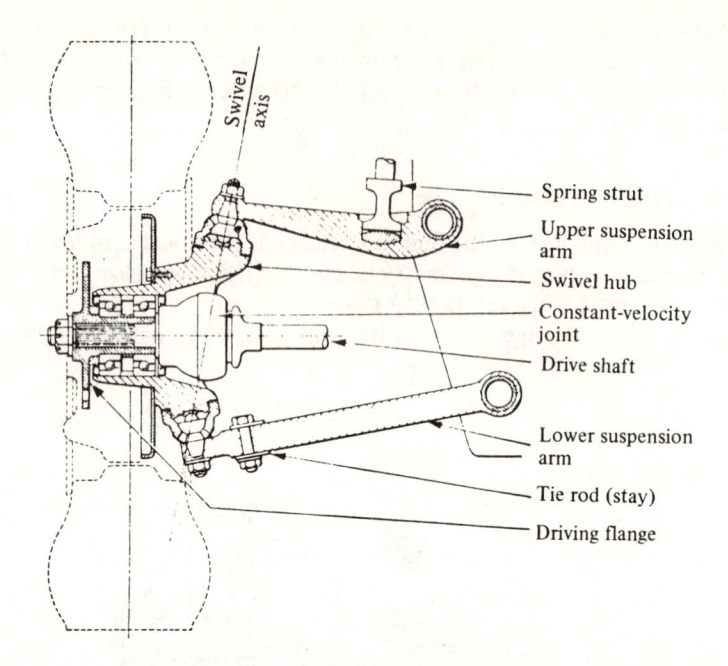

Figure 5.9 Front-wheel-drive independent front suspension

5.10 Transverse link and strut

The MacPherson arrangement combines a sliding action with a wishbone lower support (Figure 5.10). A telescopic strut is used, the lower member – forming the stub-axle assembly – sliding and turning on the upper rod. An integral hydraulic damper and coaxial road spring are used. The upper end of the strut is usually rubber mounted in a wheel arch housing and the lower end pivoted on a wishbone, one arm of which is often formed by the anti-roll bar. The steering axis of the suspension, passing through the lower ball swivel and the upper strut mounting, may not coincide with the axis of the strut.

Elimination of the upper link in favour of a strut spreads the attachment loading more widely into the body structure and gives additional space for an engine-transaxle unit. Side loading on the strut from driving, braking and cornering forces, however, requires a large-diameter piston rod; in addition, the sliding friction from its guidance role is a fundamental disadvantage.

The front-wheel assembly

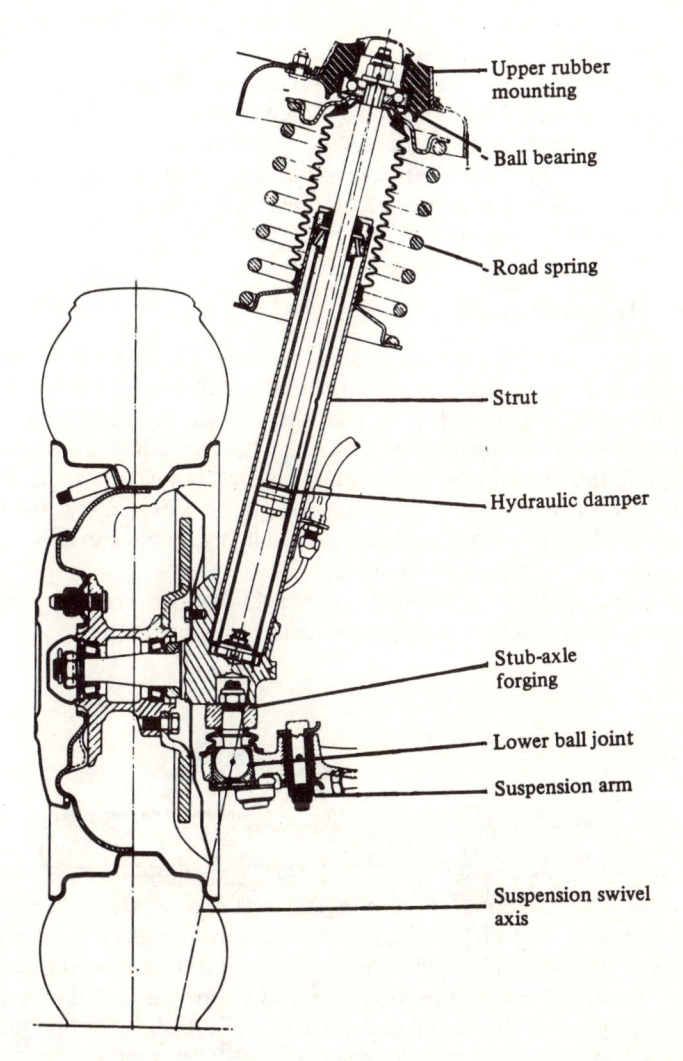

Figure 5.10 Transverse link-and-strut independent front suspension

5.11 Steering box

To reduce the driver's effort with unassisted steering, a gear reduction is provided by the steering box. With an angular rotation of the stub axle of 70–80° and a steering wheel rotation of 3–4.5 turns, a ratio of 14:1 to 23:1 is required. Heavier vehicles or vehicles with FWD engine-transaxle units require the greater leverage; alternatively, they may use power assistance with a reduced ratio.

Some degree of irreversibility or reverse inefficiency in the mechanism tends to prevent the steering wheel being rotated by kick-back moments applied to the stub axles. Legislation in some countries requires a steering column to have a sliding or collapsible portion or to be in out-of-line parts connected by universal joints so that a collision cannot impale the driver.

5.12 Rack and pinion

The most usual system on cars is a helically toothed pinion on the steering shaft operating a rack connected by ball-jointed tie rods to the steering arms (Figure 5.11). In order to obtain an adequate steering ratio – i.e. sufficient turns of the steering wheel to reduce the torque – a very small radius pinion is necessary. By using a large helix angle (40°+) on the pinion teeth and by angling the pinion axis at 20°+ from the perpendicular to the rack, a pinion of sufficient strength can be obtained with very few teeth (five or even four).

Damping of the rack results from a spring-loaded pressure pad and from the sliding friction between the heavily angled teeth.

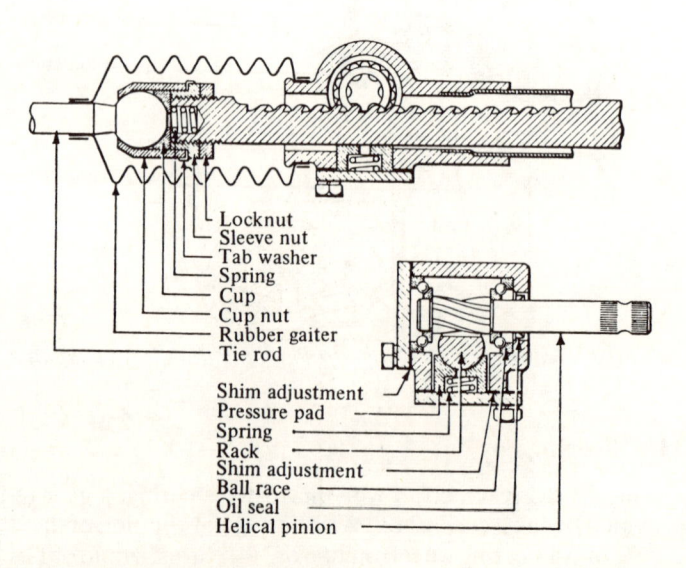

Figure 5.11 Rack-and-pinion steering box

5.13 Worm and recirculating-ball nut

This is a worm-and-nut steering box with a closed chain of ball bearings interposed between grooves in the worm and the nut and recirculating through an external tube joining the ends of these tracks. Rack teeth on the nut can mesh with integral pinion teeth on the rocker shaft. Alternatively a roller on the nut with a forked level on the shaft, or a ball on the nut with a socket arm on the shaft, can convert the axial to arcuate (arched) movement (Figure 5.12).

Another alternative occasionally used is an hour-glass worm on the steering shaft, meshing with a double or triple roller carried by the rocker arm (Figure 5.13).

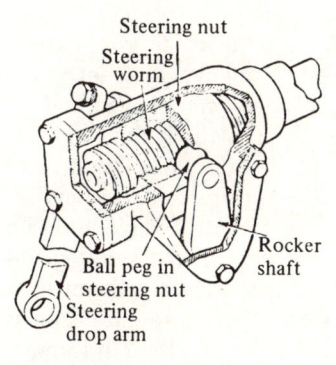

Figure 5.12 Worm-and-nut steering box

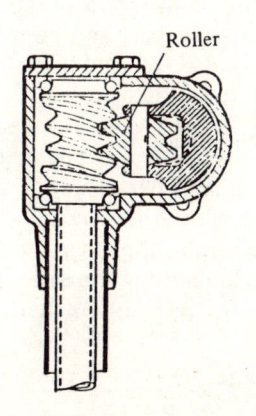

Figure 5.13 Hour-glass worm and double-roller steering box

Maintenance

The steering shaft is usually supported on ball or taper-roller bearings with a shim or screwed adjustment to eliminate end-float or to apply the specified pre-loading.

The rocker shaft normally turns in bushes and has a screwed or shim adjustment to control the end-float or the meshing adjustment. The steering box requires topping up with lubricant every 10 000 km according to type. This must be introduced via the tie-rod flexible bellows on many rack-and-pinion steering boxes; neglect often leads to premature wear.

5.14 Power-assisted steering

On an assisted system, should power fail, the driver can steer manually using increased torque. In the usual system a hydraulic pump, driven by a belt from the crankshaft pulley, has an integral reservoir with filter and pressure regulating valve and can deliver fluid at some 7–14 MPa.

Fluid pressure can be applied via the control valve to one side or the other of a double-acting hydraulic cylinder, whilst the fluid displaced from the opposite side returns to the reservoir. High-pressure hoses connect the pump to the control valve and ram. Early arrangements used a separate ram acting on the steering linkage, but modern car systems incorporate the hydraulic cylinder in the steering box. With rack-and-pinion systems the hydraulic piston is carried by the rack, whilst on recirculating-ball gears the nut forms the piston. The control system normally utilizes a torsion bar connecting the steering mast to the steering-box pinion or worm shaft. With no applied torque, fluid circulates through the control valve and back to the reservoir. Increasing torque twists the bar, and the relative movement between the steering mast and the steering-box shaft opens hydraulic ports proportionately, admitting fluid pressure to one side of the piston – depending on the direction of rotation – and releasing fluid from the opposite side. Movement of the piston restores the valve to the equilibrium position until further torque is applied by the steering wheel. By a suitable choice of torsion bar the degree of power assistance can be matched to the torque applied at the steering wheel. Torsion-bar failure is safeguarded by limiting the total relative movement between the steering mast and the steering-box shaft.

5.15 Steering linkage

With the beam axle, the drop arm on the rocker shaft is connected to one of the stub axles by a drag link or pull-and-push rod, and an adjustable track rod couples the two track arms together, ball joints being employed at the connections. The track rod and ball joints are threaded right hand on one end and left hand on the other; rotation thus increases the distance between the ball joints to adjust wheel alignment.

IFS requires a linkage arranged so that, as far as practicable, suspension movement does not affect the steering. On a rack-and-pinion layout the track adjustment is by equal rotation of both the track tie bars – right-hand threads. Other steering boxes on IFS may require an idler and multipiece connections.

Current steering-linkage ball joints use a simplified construction where a self-lubricating plastic material, compressed into place around the ball, relies on elasticity to eliminate the spring loading previously employed. Some heavily loaded suspension ball joints use a two-piece seating in order to separate and reduce the friction in the swivelling/ steering function from the suspension oscillations.

Steering geometry should not normally vary but may do so through accidental damage or wear. Regular checking is necessary, particularly after any impact, otherwise unsatisfactory steering and abnormal wear of the tyres, steering or suspension parts may occur.

Inaccuracies following an impact usually indicate bending and must not be rectified by adjustment until a complete check of the steering geometry has been made. For example, if a steering arm is bent, correction of the alignment by adjustment of the track will not give the correct Ackerman action; this fault would be indicated by incorrect cornering wheel relationship and would cause tyre scrub.

Castor, camber and KPI are usually measured by spirit-level type gauges; wheel alignment by telescopic, drag-plate or optical gauges; and cornering wheel relationship or toe-out on turn by graduated turntables. However, optical methods are available for all the tests.

The results can be nullified by wheel run-out, incorrect tyre pressures, uneven ground or, for IFS systems, an incorrect loading or setting height for the vehicle.

5.16 Wheel assembly

The front hub may revolve on two non-adjustable ball-journal bearings, in which case the securing nut must be fully tightened to clamp the two inner races and the distance piece together. If adjustable double taper-roller bearings are employed (Figure 5.14), great care is necessary not to overtighten the adjusting nut, since the wheel will revolve 'freely' even when an axial load of damaging value is applied. The best setting is a free-running clearance with no appreciable end-play – but as this is

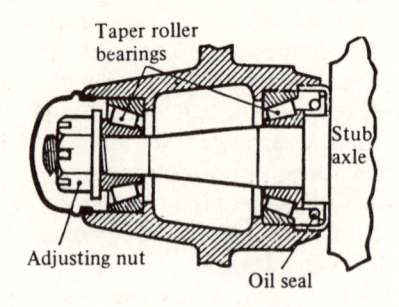

Figure 5.14 Double taper-roller bearings

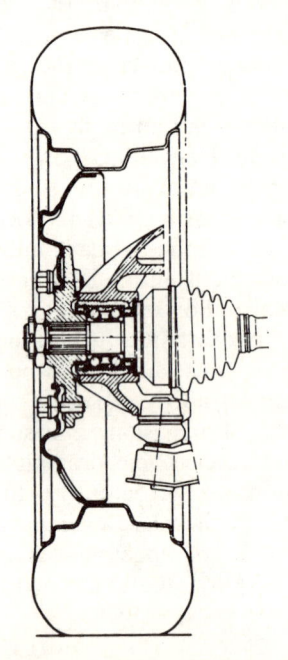

Figure 5.15 Double angular-contact ball race

difficult to ensure, an end-play of about 0.1 mm is customary with taper-roller front-wheel bearings.

Many double taper-roller and double angular-contact ball races (Figure 5.15) are now manufactured with abutment of the races, eliminating adjustment.

The hubs need packing with a limited quantity of grease at some 50 000 km intervals after dismantling and examination. An excess of grease may cause heating and loss of lubricant.

5.17 Tyres and tubes

Tyres provide the necessary frictional contact between the road and the wheels required for tractive, braking and cornering forces, and they assist in the absorbing of road shocks.

The carcase of the tyre consists of a foundation of bead wires and layers of parallel rayon or nylon cords encased in rubber compound (Figure 5.16). In a cross-ply tyre each layer makes an alternate equal angle with the circumferential centre line of between 25° and 40°, stiffness increasing with the angle. In the radial construction the few layers of cord plies are at 90° to the circumference and have a bracing belt – or breaker strip – of steel or fabric cords under the tread. The adjacent layers of this belt are angled at between 10° and 30° to the centre line. The bias-belted construction has carcase cords as in a

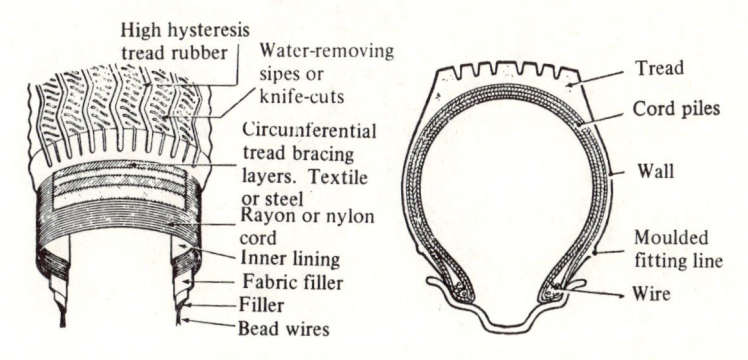

Figure 5.16   Radial (left) and cross-ply (right) tyres

cross-ply tyre plus a fabric belt, but with large-angled plies only some 5° less than those in the carcase.

Compared with cross-ply, the radial combination of flexible walls and braced tread provides increased cornering power and greater wear resistance; road noise is reduced and road impacts are better absorbed. However, low-speed steering can be heavier and there is greater vulnerability to sidewall damage. The bias-belted tyre has intermediate characteristics.

The tread and the walls are vulcanized – a process of heating the rubber under pressure – to the foundation. Special mixes and patterns – formed by the mould during vulcanizing – are used to give the tyre the characteristics suited to its use. The tread pattern must pass water rapidly across the contact area to prevent reduced grip and eventual aquaplaning on wet roads.

The tube of butyl rubber is inflated via a spring-loaded valve (Figure 5.17) to a pressure of some 100–700 kPa, depending upon the vehicle and the conditions of service. The temperature of the tyre rises during running and this results in an increased pressure, but under these conditions the excess pressure normally should not be released.

The tubeless tyre has the inner surface of the cover lined with soft rubber which extends around the bead to form an air-tight seal onto the rim. This tends to make the tyre self-sealing around small penetrating objects.

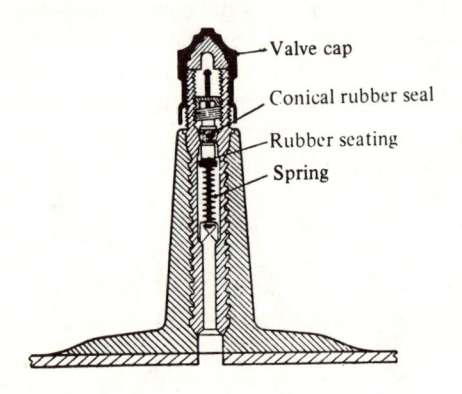

Figure 5.17   Tube valve

5.18   Tyre nomenclature

The height of the tyre section to its width – originally about equal – has been progressively reduced in the search for greater cornering power. This aspect or profile ratio is given as a percentage in the 70, 60 and 50 series low-profile tyres.

The first identification figures on the tyre wall give the section width, in millimetres for radial and in inches for cross-ply construction. This may be followed by the percentage double figure of the low-profile series. Code letters then indicate the speed rating: SR, HR and VR for radial tyres permit maximum speeds of 180, 210 and over 210 km/h, respectively; S and H on cross-ply tyres have slightly lower maximum speeds. Finally, the rim diameter is given in inches. Thus, for example, 200/60 HR 15 is a radial tyre of 200 mm nominal section width, a profile ratio of 60 per cent (i.e. a nominal section height of 120 mm), limited to a maximum speed of 210 km/h, and fitting a 15 in rim diameter.

Commercial tyres may include a number followed by PR which refers to the ply rating. This relates to the load-carrying capacity of the tyre – not necessarily the number of plies in the carcase.

5.19 Tyre usage

Cars should preferably be fitted with the same type of tyre all round. This permits interchanging and retains the original handling characteristics. Cross-ply and radial tyres must not be used on the same axle and, if the types are mixed, the radials must be on the rear to prevent oversteer. Similarly, should textile and steel-braced radial tyres be mixed, the latter should be fitted to the rear axle.

The factors promoting maximum tyre mileage are correct tyre pressures, balanced wheels, accurate camber angle and wheel alignment, correct brake action, systematic changing of the position and direction of rotation of the tyres every 10 000 km, and the avoidance of abnormal acceleration, speed or braking.

Typical legislation for water dissipation from the contact area requires a continuous pattern across the full width of the tread, having a minimum depth of 1 mm.

Tubeless tyre fitting, though easier than with a tubed tyre, requires care to avoid bead damage and to ensure a satisfactory rim seal. The edges of a tyre contain an inextensible wire ring; thus where a well-base rim is employed, one portion of the bead must be pressed to the well before the other side can be removed (Figure 5.18). The valve portion should be removed first and replaced last. The bead seatings in the rim have a taper of 5–15°, and after a period in position considerable force

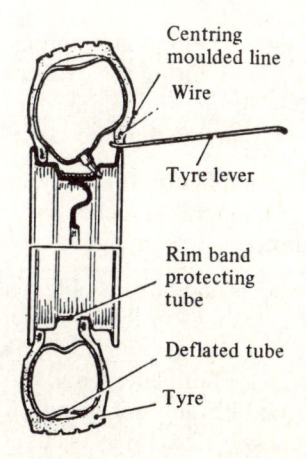

Figure 5.18   Tyre removal

may be needed to free the beads. This is conveniently applied by powered tyre equipment; never attempt to hammer a tyre.

Methods of providing a safe run-flat facility for well-base rims include modified rim sections or well-base fillers to prevent the tyre leaving the rim after a high-speed deflection.

The more rigid tyres used by commercial vehicles cannot be deformed for fitting, and the outer flange of the rim is made detachable. In the two-piece type, a split flange is sprung into a specially formed groove in the edge of the rim base. In the three-piece type, a continuous flange is secured by a split locking ring (Figure 5.19).

Two-piece rim
Detachable spring flange
Tapered bead-seat surface

Three-piece rim
Continuous flange
Spring lock-ring

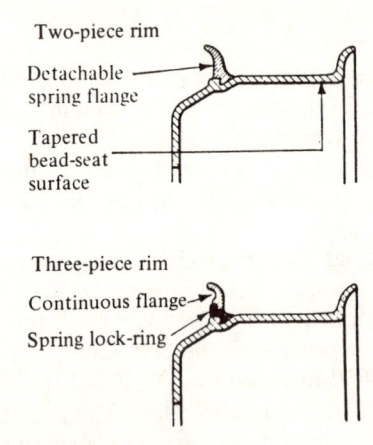

Figure 5.19 Wide-base commercial rims

Care must be taken before attempting removal that the tyre is completely deflated, and during removal that the split flange or ring is not distorted. When fitting, the mating faces must be clean and must seat perfectly. Initial inflation must be carried out with the wheel inside a steel grille.

5.20 Wheel balance

Concentricity of wheels and hubs is obtained by the wheel nuts having centralizing conical seatings in the wheel pressing or, less frequently, from the fit of the wheel centre on the hub spigot. The standard tolerance for car wheels – for both lateral and radial variation – is 2 mm.

If the wheel assemblies are not balanced, vibration or rapid wheel oscillation about the swivel axes may occur at certain road speeds. Static balancing entails attaching a balance mass to the wheel rim such that the

The front-wheel assembly

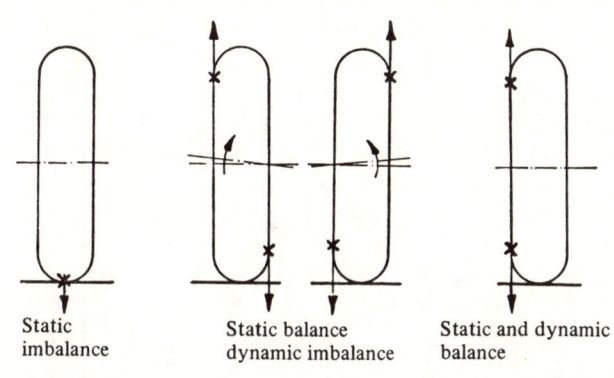

Static imbalance | Static balance dynamic imbalance | Static and dynamic balance

Figure 5.20   Wheel balance

assembly will rest in any position. Unless this balance mass is in the same wheel plane as the original out-of-balance force, there will be dynamic imbalance during rotation, with a rocking couple (Figure 5.20).

5.21   Diagnosis of tyre wear

Appearance
Even circumferential wear
One side of tread worn
Centre of tread worn
Shoulders of tread worn
Feather edges on edges of the tread pattern:
    On inside edges on both tyres
    On outside edges on both tyres
    On nearside tyre only

Uneven circumferential wear
Severe scalloping
Uneven wear

Indication

Excessive camber angle
Over inflated
Under inflated

Excessive toe-in
Excessive toe-out
Effect of road camber

Oval brake drums. Brake judder
Worn suspension or worn wheel bearing
Tyre pressure or alignment faulty
Unbalanced wheel
Irregular brake action

Example 10

A vehicle has a wheelbase of 3.360 m and a track of 1.320 m (Figure 5.21). Determine the angles through which the inner and outer front

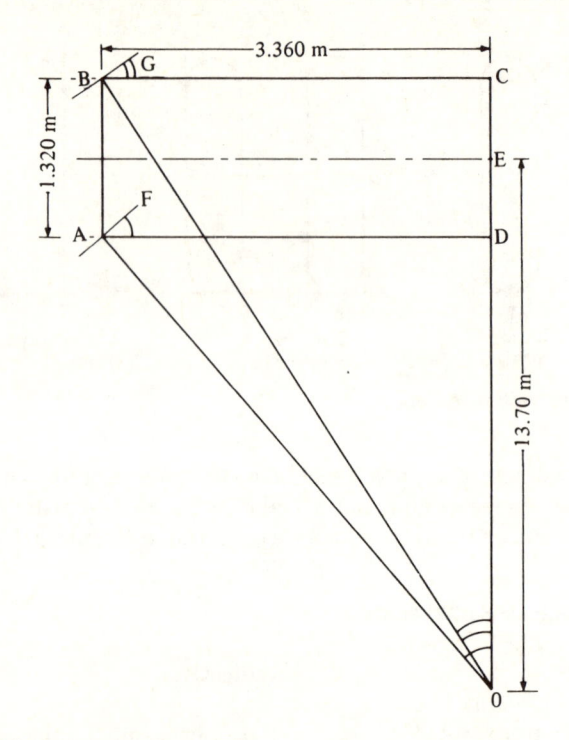

Figure 5.21

wheels must be turned in order that the vehicle may corner about a point 13.70 m from its centre line.

Let AB and CED be the front and rear axles of the vehicle. Then the centre of turning must lie at O on CED produced such that EO = 13.7 m. Now $\angle$ FAD = angle turned through by the inner wheel. But $\angle$ FAO = 90°, since AO is the produced wheel axis. Hence

$$\angle FAD = 90° - \angle DAO$$

In triangle ADO, $\angle$ ADO = 90°, therefore $\angle$ DOA + $\angle$ DAO = 90°, since interior angles together equal 180°. Hence

$$\angle DOA = 90° - \angle DAO$$

Hence

$\angle$ DOA = $\angle$ FAD = angle turned through by inner wheel

The front-wheel assembly

Now

$$\tan \angle DOA = \frac{\text{opposite}}{\text{adjacent}}$$

$$= \frac{AD}{DO}$$

$$= \frac{3.360}{13.70 - 0.660}$$

$$= 0.2577$$

$$= \tan 14°$$

Similarly,

$\angle COB = \angle GBC$ = angle turned through by outer wheel

and

$$\tan \angle COB = \frac{CB}{CO}$$

$$= \frac{3.360}{13.70 + 0.660}$$

$$= 0.2340$$

$$= \tan 13°$$

The angles turned through by the inner and outer wheels are therefore 14° and 13° (to the nearest degree).

Chapter 6
Engine principles

6.1 The four-stroke cycle

Induction stroke

On modern high-speed units, to obtain the maximum filling of the cylinder the inlet valve opens about 10° before top dead centre (TDC), while the exhaust valve closes some 10° after TDC, giving 20° of valve overlap (Figure 6.1). During this period the momentum of the outgoing exhaust gas creates a partial vacuum within the cylinder even before the piston has moved far on the induction stroke. The atmospheric pressure, being greater than the depression in the cylinder, causes a flow of vaporized fuel and air from the carburettor. The inlet valve remains open until some 50° after bottom dead centre (BDC) to take advantage of the momentum of incoming mixture, which may have a speed of 30–90 m/s.

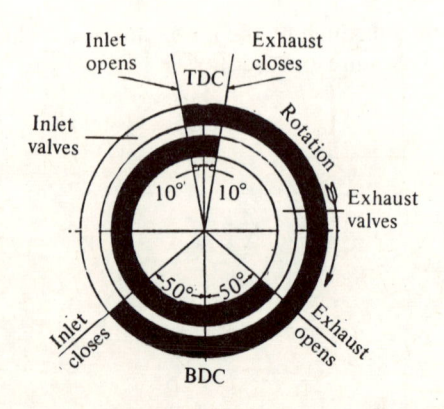

Figure 6.1 Typical valve timing diagram

Engine principles

Compression stroke

On the petrol engine the charge is compressed into about one-eighth or one-ninth of its original volume. The pressure rises to about 1 MPa, depending upon factors including the compression ratio, throttle opening and engine speed. Near the top of the stroke the mixture is ignited by a spark which bridges the gap at the plug points.

Power stroke

Combustion of the charge results in very high temperatures, which may exceed 2000°C on full throttle. The high temperature produces a rise in pressure of the gas to some 3.5 MPa, followed by a fall in pressure as the piston traverses the power stroke.

Exhaust stroke

The exhaust valve opens some 50° before BDC, allowing the pressure within the cylinder to fall and so reducing back pressure on the piston during the exhaust stroke. The valve remains open until about 10° after TDC.

6.2 The two-stroke cycle

We shall consider the three-port, crankcase-compression, single-cylinder engine (Figure 6.2).

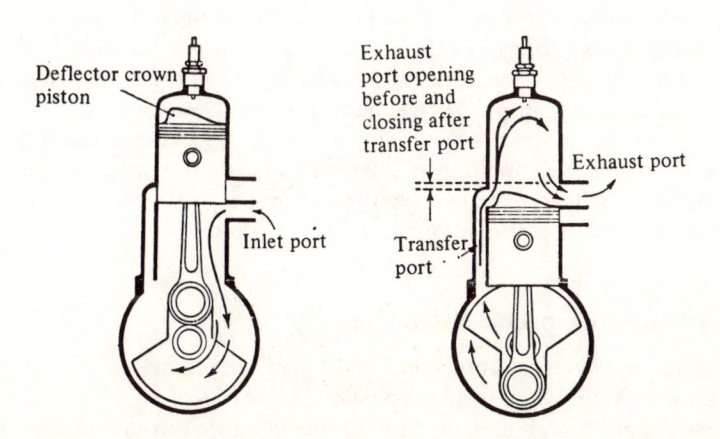

Figure 6.2 Three-port two-stroke engine

Events above the piston

The piston rises, compressing the charge in the combustion chamber. Near TDC the compressed gas is ignited by a spark bridging the gap at the plug points. The high temperature produced by the combustion of the charge causes a rise in pressure, followed by expansion as the piston travels down.

Near BDC there are two ports in the cylinder wall – the transfer port, communicating with the crankcase, and the exhaust port. As the piston nears BDC it uncovers first the exhaust port, releasing the burnt gas, and then the transfer port, admitting to the cylinder a compressed charge from the crankcase.

Events below the piston

Each upward stroke of the piston creates a partial vacuum in the crankcase, and near TDC the lower edge of the piston skirt uncovers the inlet port, admitting a fresh charge of gas from the carburettor.

As the piston descends, the inlet port is closed and the charge is compressed in the crankcase until the transfer port is exposed, allowing the gas to escape into the cylinder.

A deflector-head piston is used to direct the incoming gas to the top of the cylinder. The crankcase must be small and gas-tight to provide adequate crankcase compression, and the piston rings must be pegged to prevent rotation and their ends being trapped in a port. Lubrication is by mixing some 50 ml of oil with each litre of petrol (a 'petroil' mixture) or by metering oil from a throttle-regulated oil pump. The simplicity, low weight and more even torque of the engine makes it popular for low-powered transport, small industrial purposes and portable tools. Alternatively, developed porting and exhaust design, with the solution of heat dissipation and distortion problems, enable an unrivalled power output to be achieved in specific capacities.

The unavoidable loss of fresh mixture through the exhaust ports, which must close after the transfer in the orthodox layout, and contamination with residual burned gas, result in poor economy and in emissions unacceptable for general transport use.

6.3 Indicated pressure and power

The actual pressure within the cylinder of an engine varies and does useful work only on the power stroke.

By means of an indicator, this varying pressure throughout the cycle of operations can be recorded on an *indicator diagram* (Figure 6.3).

Engine principles

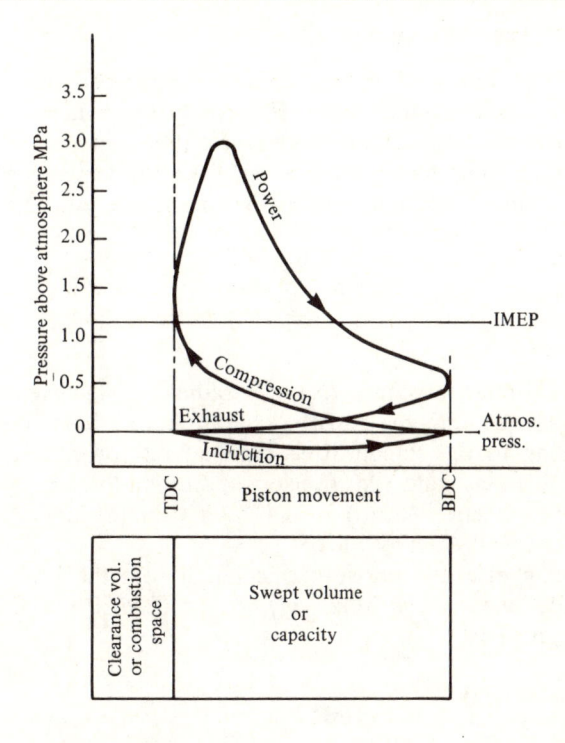

Figure 6.3 Indicator diagram

From the indicator diagram the average pressure which is effective in doing useful work, or the *indicated mean effective pressure* (IMEP), can be calculated.

The product of the IMEP and the piston area gives the effective force on the piston, and this, multiplied by the stroke, gives the work done per stroke. Hence the *indicated power* (IP: watts), the power developed inside the cylinder, is given by

$$IP = IMEP \times lan$$

where IMEP is in pascals (Pa), l = stroke (m), a = piston area (m²), and n = number of effective strokes per second, where

four-stroke: $\quad n = \dfrac{\text{rev/s}}{2} \times$ number cylinders

two-stroke: $\quad n = \text{rev/s} \times$ number cylinders

6.4 Brake pressure and power

About 15 per cent of the indicated power is absorbed by frictional and pumping losses within the engine. The remaining useful power available at the flywheel is termed the *brake power* (BP: watts), since it can be measured by some form of frictional, hydraulic or electrical 'brake' driven from the crankshaft. The BP in watts can be calculated from the formula:

$$BP = 2\pi Frn$$
$$= 2\pi Tn$$

where F = brake resistance (N), r = radius at which resistance acts (m), T = torque Fr (N m), and n = crankshaft rotational speed (rev/s).

According to the standardized national test procedure used, the engine power may state SAE (Society of Automotive Engineers), DIN (Deutsche Industrie Norm) or CUNA (Commissione Technica di Unificazione Nell'Autoveicolo).

The mean effective pressure calculated from the brake power is termed the *brake mean effective pressure* (BMEP). The BMEP in pascals is given by

$$BMEP = \frac{BP}{lan}$$

where BP is in watts (W), l = stroke (m), a = piston area (m^2), and n = number of effective strokes per second (calculated as in Section 6.3).

The BMEP is an indication of engine efficiency regardless of capacity or engine speed; 1000 kPa represents high efficiency.

6.5 Mechanical efficiency

Indicated power is related to brake power by

$$IP = BP + FP$$

where FP is the *frictional power*, that is the power absorbed within the engine by friction and pumping losses during the induction and exhaust strokes.

The ratio BP/IP is termed the *mechanical efficiency* (symbol η), and has a maximum of about 85 per cent. At higher engine speeds a greater proportion of the power is absorbed in friction and consequently the mechanical efficiency falls.

Engine principles

6.6 Volumetric efficiency

The ratio of the mass of charge actually induced on the piston induction stroke to the mass of charge required to completely fill the swept volume at atmospheric pressure and temperature is termed the *volumetric efficiency*.

Volumetric efficiency depends upon throttle opening and engine speed as well as induction and exhaust system layout, port sizes and valve timing. With fixed valve timing the highest volumetric efficiency can only be obtained at a specific speed, and has a maximum of about 80 per cent with unsupercharged engines. Below this speed the valve timing is not so suitable and the volumetric efficiency falls slightly, while at high engine speeds the volumetric efficiency is considerably reduced.

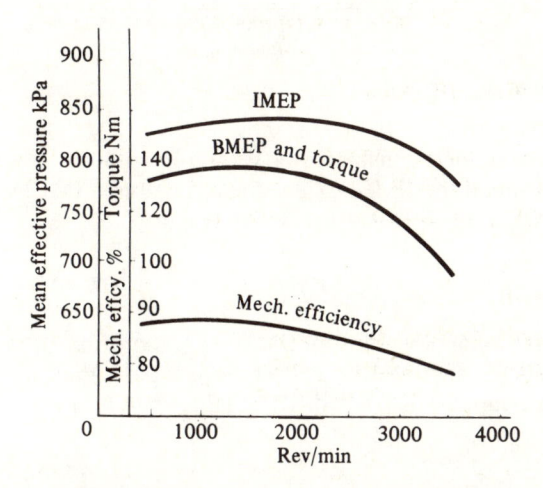

Figure 6.4   BMEP and IMEP curves: full-throttle conditions

6.7   Curves of IMEP, IP, BMEP and torque

IMEP

At higher engine speeds the IMEP falls off, chiefly owing to the lowered volumetric efficiency, since a reduced mass of charge is being burned per power stroke (Figure 6.4).

IP

The IP rises until it reaches a maximum, where a proportional increase in revolutions is balanced by a proportional reduction in the IMEP; if

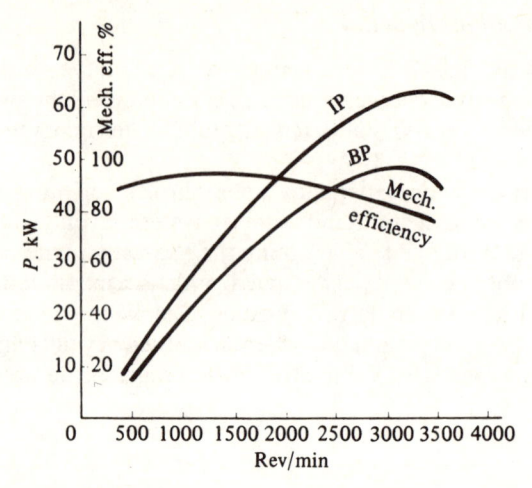

Figure 6.5 IP and BP curves

the IMEP remained constant, the IP graph would be a straight line. Above this speed the IP falls, since the reduction in IMEP is greater in proportion than the increase in rev/min (Figure 6.5).

BP and BMEP

Since the frictional power loss becomes a greater proportion of the IP as the speed increases, the BP falls off more rapidly than the IP and the BMEP more rapidly than the IMEP (Figures 6.4, 6.5).

Torque

Since the BMEP represents the average force on the piston producing torque at the flywheel, BMEP and torque are directly proportional and the same curve with a different vertical scale can be used for each.

Engine curves are always given for full throttle test conditions, which rarely prevail in road use.

6.8 Thermal efficiency

Work can be completely converted into heat by friction, but the reverse process is not possible. The internal combustion engine can convert only about one-quarter of the heat energy supplied by the fuel into work; the remaining three-quarters is dissipated, mainly by the exhaust gases and

Engine principles

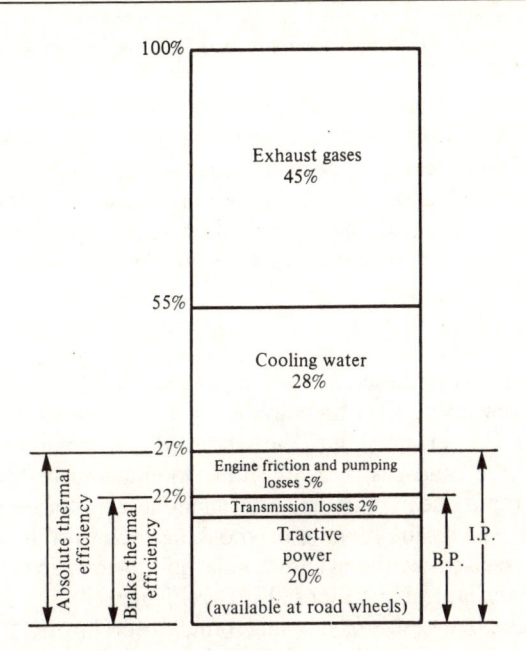

Figure 6.6 Distribution of fuel energy

the cooling system (Figure 6.6). The ratio of work done to heat supplied is called the *thermal efficiency*. Depending upon whether the work done is measured at the flywheel by a 'brake' or in the cylinder, the ratio is given as the brake thermal efficiency or the absolute thermal efficiency.

6.9 Compression ratio

The compression ratio of an engine is the ratio of the charge volume before and after compression:

compression ratio

$$= \frac{\text{total volume with piston at BDC}}{\text{total volume with piston at TDC}}$$

$$= \frac{\text{swept volume (cylinder capacity)} + \text{clearance volume (combustion space)}}{\text{clearance volume (combustion space)}}$$

where

$$\text{swept volume} = \pi \times \left(\frac{\text{bore}}{2}\right)^2 \times \text{stroke}$$

The *bore* is the cylinder diameter, and the *stroke* is the length of the swept volume.

The thermal efficiency of an engine depends fundamentally upon the compression ratio. Raising the compression ratio improves the thermal efficiency, but the gain becomes progressively smaller as the compression ratio increases. In practice a limit is reached, depending upon the engine design and the fuel used, when any further rise in the compression ratio causes detonation.

6.10 Detonation

Under normal conditions the burning of the charge is a comparatively steady process, but if during combustion the temperature or pressure rises beyond a certain point, the whole of the remaining gas in the cylinder – the 'end-gas' – may ignite spontaneously, producing an extremely rapid rise in pressure. This phenomenon, termed *detonation*, is characterized by the *pinking* or knocking sound produced when the pressure wave strikes the cylinder walls and piston crown and creates abnormal mechanical and thermal stress (Figure 6.7).

Detonation can result from using a fuel of too low octane rating for the compression ratio of the engine, and is also influenced by the design

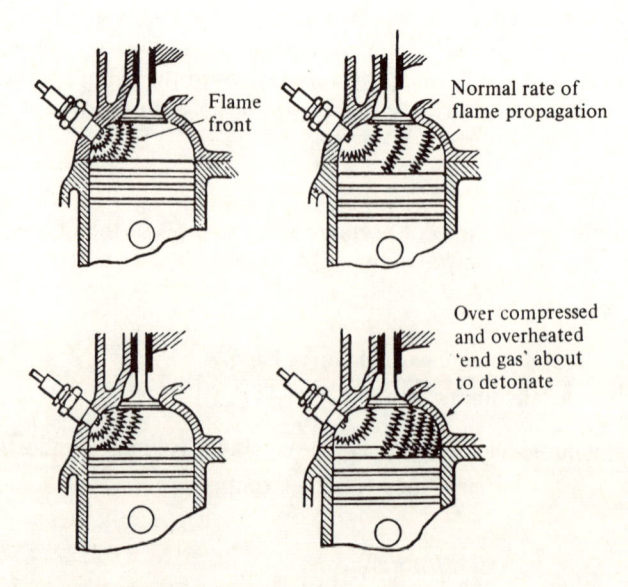

Figure 6.7   Normal combustion and detonation

of the combustion chamber. Turbulence of the charge, a short flame travel, and the movement of the flame front towards a cooler region of the combustion chamber reduce the risk of detonation. A weak mixture, high intake air temperature, an overheated engine, excessive ignition advance or carbon deposits – which will restrict cooling and raise the compression ratio – can all promote detonation.

Detonation is most pronounced when accelerating or hill climbing where, with full throttle, the engine is inducing a maximum charge at a comparatively low piston speed.

6.11 Pre-ignition

Pre-ignition occurs when the charge is ignited before the normal ignition point by some incandescence in the combustion chamber. An undesirable rise in pressure is produced before the piston reaches TDC, causing a loss of power and excessive stress in the mechanical parts.

Pre-ignition may be caused by the central electrode of an incorrect type of plug, the sharp rim of an exhaust valve, the projecting edge of a cylinder head gasket, carbon, or any other incandescent projection in the combustion chamber. Pre-ignition frequently follows detonation and is promoted by the same causes. In some cases the engine will 'run on' when the ignition is switched off.

6.12 Single- and multi-cylinder engines

Power/weight ratio

With a given piston speed and BMEP (not usually exceeding 16 m/s and 1000 kPa), the engine power varies as the square of the bore (that is, with the piston area) but the mass varies as the cube of the bore (that is, with the volume of metal used). Increasing power by using a large cylinder therefore results in a low power/weight ratio, whereas increasing the number of cylinders maintains power and weight in the same proportions.

Firing interval and torque fluctuation

Since all the cylinders must fire in two revolutions of the four-stroke crankshaft, the firing interval is 720° divided by the number of cylinders. The effective power stroke occupies about 135°. With a single cylinder the mass of a large flywheel is required to absorb torque fluctuations and provide energy for the crankshaft. As the number of cylinders increases, torque is smoother and less flywheel mass is needed, aiding acceleration.

Cooling

Large cylinders have long heat paths, such as from the piston centre. Multi-cylinder units are necessary for large power to avoid lubrication and detonation problems due to overheating.

Balance and inertia loads

The single-cylinder unit can only be imperfectly balanced and vibration will occur at certain engine speeds. Four-cylinder in-line units have small secondary out-of-balance forces, while horizontally opposed, six- and eight-cylinder units can have entirely satisfactory balance. The reduced reciprocating mass of the multi-cylinder engine permits higher crankshaft speeds without inertia force problems.

Conventional car engine

Apart from the benefits of traditional experience in this type of unit, the four-stroke, four-cylinder, in-line, water-cooled petrol engine has inherent advantages.

The two-stroke unit has unacceptable fuel consumption. The economy of the compression-ignition (CI) engine is offset by the lower power and acceleration, with increased cost, noise, weight and (to some) more objectionable fuel.

Twin-cylinder engines have greater torque fluctuations, and six-cylinder units are an unnecessary expense under 2–2.5 l capacity. The in-line layout is cheaper; fewer components are needed than for the V4 or the horizontally opposed four (HO4). The V4 usually requires a balance shaft, and the HO4 has complicated manifolds and cooling arrangements.

Air cooling is not suited to four-cylinder in-line units; it is noisier, requires power to drive the large cooling fan, and complicates interior heating.

6.13 Crankshaft layout

To obtain the best balance the four-cylinder engine employs a crankshaft arrangement in which the front and rear pistons are at TDC when the centre pistons are at BDC. The firing order must be either 1 3 4 2 or 1 2 4 3 depending upon the arrangement of the cams on the camshaft (Figure 6.8).

The six-cylinder unit has a crankshaft with the throws at 120° to each other, and the corresponding pistons at front and rear moving together.

Engine principles

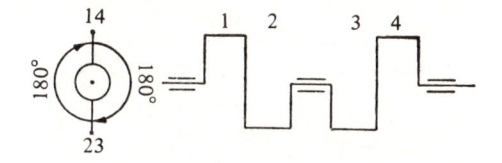

Cylinder 1 2 3 4	Firing order
P E C I E I P C I C E P C P I E	1 3 4 2

Cylinder 1 2 3 4	Firing order
P C E I E P I C I E C P C I P E	1 2 4 3

I = induction stroke
C = compression stroke

P = power stroke
E = exhaust stroke

Figure 6.8   Four-cylinder firing orders

Firing orders of 1 5 3 6 2 4 or 1 4 2 6 3 5 are employed in order to obtain a well-balanced mixture distribution by feeding alternately to the front and rear of the unit (Figure 6.9).

Three- and five-cylinder in-line engine crankshafts have the throws at 120° and 72° spacing respectively, giving firing orders of 240° and 144°.

The simplest V8 layout is, in effect, two four-cylinder units at 90° with each throw of the common crankshaft carrying a connecting rod from each bank. Superior balance is usually obtained by using a two-plane crankshaft, with front and rear throws at 180° and the central two also opposite but at right angles to the first pair. The firing interval is 90°.

6.14   Combustion chamber layout

Combustion chamber design has an important influence on BMEP, economy and emission control. A low surface/volume ratio is required to reduce heat loss and improve thermal efficiency. Efficient combustion is promoted by turbulence of the charge within the cylinder head. With suitable inlet tract design the momentum of the high-speed incoming gas will produce *swirl* turbulence continuing throughout compression. *Squish* turbulence – produced by displacing the charge at

Firing order 1·5·3·6·2·4

Firing order 1·4·2·6·3·5

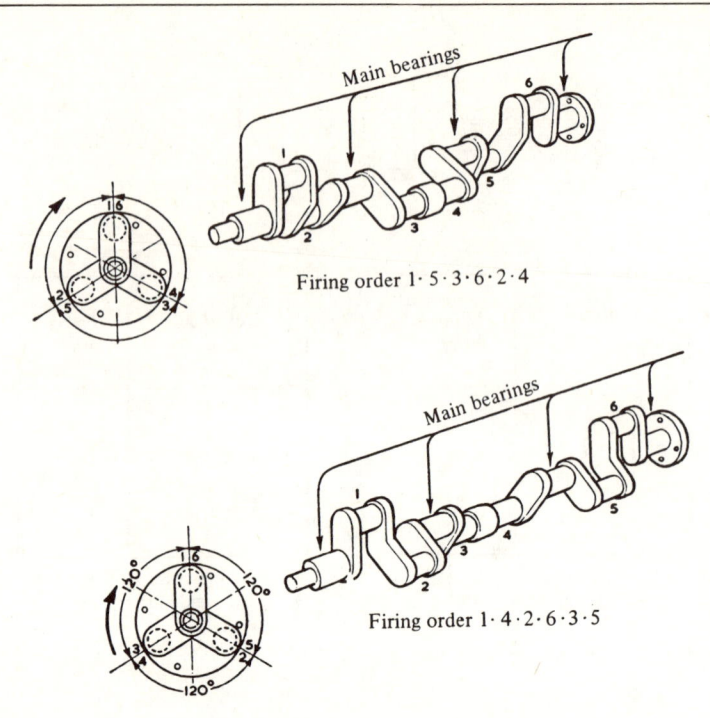

Figure 6.9 Six-cylinder crankshaft layout

high speed into a pocket during the last stages of compression – is obtained with the wedge, bath-tub or Westlake-heart chambers. All use in-line valves – the first inclined, the others vertical – operated by push rods and rockers from a side camshaft or by an overhead camshaft (Figure 6.10).

Both the hemispherical and the penthouse chambers, with inlet valves inclined from one side and exhausts from the other, offer slightly improved thermal and volumetric efficiency where complication and cost are less important. A four-valve layout increases valve area, improves valve cooling, reduces valve-spring load and facilitates a central plug position with short flame travel. Valve operation may be by two overhead camshafts or by a single camshaft and rockers. The cross-flow principle can also be advantageously employed to improve the volumetric efficiency and swirl turbulence of a vertical in-line valve layout.

In the Heron design the combustion chamber is located in the piston crown. With a flat cylinder head face and die-cast pistons the clearance volumes are precise without the necessity for head chamber machining.

Engine principles

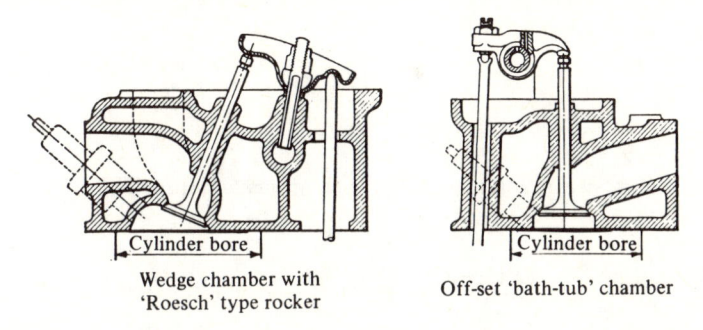

Cylinder bore — Wedge chamber with 'Roesch' type rocker

Cylinder bore — Off-set 'bath-tub' chamber

Figure 6.10 Cylinder heads with squish turbulence, short flame travel and cool end-gas

Considerable turbulence is obtained from the annular squish band, but there is some increase in piston mass and heat transfer to the piston.

The May Fireball is a fully machined split-level cylinder head chamber. The lower collecting zone below the inlet valve receives an unusually weak mixture; this is displaced by the rising flat-top piston along a tangential guide ramp into the upper combustion zone, below the exhaust valve, where it is ignited by a high-energy spark. The plug receives the fresh charge but is shielded by a small pocket from the swirling gas to allow the flame nucleus to develop. The characteristics of the chamber – with induced swirl and an unusually high compression ratio – together with fuel injection and high-energy ignition, enable high thermal efficiency and BMEP to be combined with low emission. The cooling from the excess air reduces peak temperature, minimizing oxides of nitrogen, and the reduced piston clearances also help to minimize unburned hydrocarbons (see Section 11.12).

Example 11

An engine has a swept volume of 240.3 cm^3. What will be the clearance volume to give a compression ratio of 8.4 to 1?

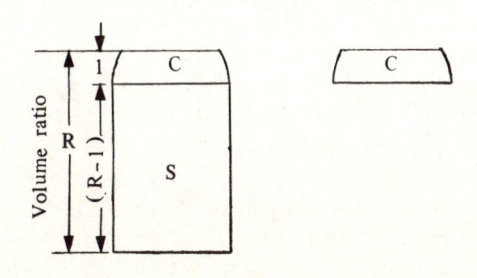

Figure 6.11

The compression ratio ($R/1$ or R) is given by (Figure 6.11):

$$\frac{R}{1} = \frac{S+C}{C}$$

where S = swept volume and C = clearance volume. Therefore

$$RC = S + C$$
$$(RC) - C = S$$
$$C(R-1) = S$$

$$C = \frac{S}{R-1}$$

In words,

$$\text{clearance volume} = \frac{\text{swept volume}}{(\text{compression ratio} - 1)}$$

Hence

$$= \frac{240.3}{(8.4 - 1)}$$

$$= \frac{240.3}{7.4}$$

$$= 32.5 \text{ cm}^3$$

Example 12

An engine has an original bore of 86 mm and a stroke of 72 mm. It is rebored 1.5 mm oversize. Calculate the original capacity, the increase and the percentage increase in capacity. The original compression ratio is 8.2:1. Calculate the compression ratio after reboring and the percentage increase.

$$\text{capacity in cm}^3 = \frac{\pi}{4} d_i^2 h$$

$$= \frac{\pi}{4} \times 8.6^2 \times 7.2$$

$$= 418.23 \text{ cm}^3$$

where d_1 = original bore in cm, d_2 = rebored bore in cm and h = stroke in cm.

$$\text{increase in capacity} = \frac{\pi}{4} h (d_2^2 - d_1^2)$$

$$= \frac{\pi}{4} h (d_2 - d_1)(d_2 + d_1)$$

$$= \frac{\pi}{4} \times 7.2 \times 0.15 \times 17.35$$

$$= 14.72 \text{ cm}^3$$

Therefore

$$\% \text{ increase} = \frac{14.72}{418.23} \times 100$$

$$= 3.52\%$$

Now, as before,

$$\text{clearance volume} = \frac{\text{swept volume}}{\text{compression ratio} - 1}$$

$$= \frac{418.23}{7.2}$$

$$= 58.09 \text{ cm}^3$$

Therefore

$$\text{final compression ratio} = \frac{\text{swept volume} + \text{clearance volume}}{\text{clearance volume}}$$

$$= \frac{(418.23 + 14.72) + 58.09}{58.09}$$

$$= 8.45$$

and

$$\% \text{ increase} = \frac{0.25}{8.2} \times 100$$

$$= 3.05\%$$

Example 13

A six-cylinder four-stroke engine running at 4000 rev/min develops a BP of 45 kW. Calculate the BMEP if the bore is 70 mm and the stroke 68 mm.

We have, for BP in watts and BMEP in pascals,

$$\text{BMEP} = \frac{\text{BP}}{lan}$$

where

l = length of stroke (0.068 m)
a = piston area ($\pi \times 0.035 \times 0.035$ m^2)
n = number of power strokes per second ($[4000/60]/2 \times 6 = 200$)

Hence:

$$\text{BMEP} = \frac{45\,000}{0.068 \times \pi \times 0.035 \times 0.035 \times 200}$$

$$= 859\,781 \text{ Pa} = 860 \text{ kPa}$$

Example 14

A four-cylinder engine has a total swept volume of 0.984 litres and develops a BP of 22.4 kW at 3000 rev/min. Calculate the BMEP.

As before

$$\text{BMEP} = \frac{\text{BP}}{lan}$$

where

l = length of stroke (m)
a = piston area (m^2)
n = number of power strokes per second ($[(3000/60)/2] \times 4 = 100$)

The swept volume or capacity is given for all four cylinders. Now

$$\text{capacity of one cylinder (in m}^3\text{)} = \text{stroke (m)} \times \text{piston area (m}^2\text{)}$$
$$= la$$

Hence

$$la = \frac{\text{total swept volume}}{4} \, m^3$$

$$= \frac{0.000\,984}{4}$$

$$= 0.000\,246 \, m^3$$

Therefore

$$\text{BMEP} = \frac{22\,400}{0.000\,246 \times 100}$$

$$= 910\,569 \, N/m^2$$

$$= 910 \, kPa$$

Example 15

A four-cylinder four-stroke engine of bore 88 mm and stroke 101 mm overcomes a resistance of 214 N at a radius of 0.550 m when running at 4.230 rev/min. Calculate the BP and BMEP of the engine.

We have

$$\text{BP} = 2\pi F r n$$

where

F = brake resistance (214 N)
r = radius at which F acts (0.550 m)
n = rotational speed (70.5 rev/s)

Hence

$$\text{BP} = 2 \times \pi \times 214 \times 0.55 \times 70.5$$
$$= 52\,137 \, W$$

Now

l = stroke (0.101 m)
a = piston area ($\pi \times 0.044 \times 0.044 \, m^2$)
n = number of power strokes per second ($(70.5/2) \times 4 = 141$)

As before,

$$\text{BMEP} = \frac{\text{BP}}{lan}$$

$$= \frac{52\,137}{0.101 \times \pi \times 0.044 \times 0.044 \times 141}$$

$$= 601\,936 \text{ Pa}$$

$$= 602 \text{ kPa}$$

Example 16

An engine develops maximum brake power of 41 kW at 3480 rev/min and maximum torque of 124 Nm at 2280 rev/min. Calculate the percentage of maximum BP developed at maximum torque and the percentage of maximum torque developed at maximum power.

First,
$$\text{BP developed at maximum torque} = 2\pi Frn$$

where:

$$Fr = \text{maximum torque (124 Nm)}$$
$$n = \text{rev/s at maximum torque (38 rev/s)}$$

Hence:

$$\text{BP developed at maximum torque} = 2 \times \pi \times 124 \times 38$$

$$= 29\,606 \text{ W}$$

$$\text{percentage of maximum BP} = \frac{29\,606}{41\,000} \times 100$$

$$= 72\%$$

Now

$$\text{maximum BP} = 2\pi Frn$$

where:

$$Fr = \text{torque at maximum BP (Nm)}$$
$$n = \text{rev/s at maximum BP (58 rev/s)}$$

Hence:

$$\text{torque produced at maximum BP} = \frac{\text{maximum BP}}{2\pi n}$$

$$= \frac{41\,000}{2 \times \pi \times 58}$$

$$= 112.5\,\text{Nm}$$

Therefore

$$\text{percentage of maximum torque} = \frac{112.5}{124} \times 100$$

$$= 91\%$$

Example 17

In the Morse test of a four-cylinder engine the BP was first determined with all the cylinders firing and was found to be 26.6 kW. The BP was then determined while each cylinder was shorted out in turn, and 18.8, 18.7, 18.8 and 18.5 kW were produced with nos 1, 2, 3 and 4 cylinders shorted out respectively. Calculate the IP and the mechanical efficiency.

When one cylinder is shorted out, the total BP is reduced by the BP of the shorted cylinder and the frictional power required to 'motor' this cylinder. Hence

BP developed with one cylinder shorted out
= BP of four cylinders − (BP + FP of shorted cylinder)
= BP of four cylinders − IP of shorted cylinder

Or:

IP of shorted cylinder
= BP of four cylinders − BP developed with one cylinder shorted out

Therefore:

IP of no. 1 cylinder = 26.6 − 18.8 = 7.8
IP of no. 2 cylinder = 26.6 − 18.7 = 7.9
IP of no. 3 cylinder = 26.6 − 18.8 = 7.8
IP of no. 4 cylinder = 26.6 − 18.5 = 8.1

and
$$\text{total IP} = 31.6 \text{ kW}$$

Then
$$\text{mechanical efficiency} = \frac{\text{BP}}{\text{IP}}$$
$$= \frac{26.6}{31.6} \times 100\%$$
$$= 84\%$$

The IP of the engine is 31.6 kW and the mechanical efficiency is 84%.

Chapter 7
Engine components

7.1 Cylinder block and crankcase

The most common form is a close-grained casting in grey iron comprising cylinder block and crankcase, internally webbed to carry the main and camshaft bearings and sometimes extended below the crankshaft centre line for maximum rigidity. The cylinder bore walls are usually surrounded by water but adjacent walls are sometimes joined or 'siamesed' – often in the redesign of an existing block to a larger capacity – for the economical use of the same transfer machining line.

At the rear an integral flange can be used to give good alignment and beam strength of the block-gearbox unit. Alternatively a bolted-on engine bearer plate may be used to attach the clutch-gearbox casting or the clutch bell housing.

Occasionally, when weight-saving considerations override cost – as on some compression-ignition units – a linered aluminium-alloy cylinder block may be used. Alternatively the cylinder casting may be in iron or linered aluminium alloy with a separate crankcase in aluminium or magnesium alloy.

The cylinder head may be cast iron or aluminium alloy. The latter permits a rise of about 0.5 in the compression ratio owing to its improved conductivity, but requires inserted valve seats and guides.

The sump is usually a steel pressing but may be a light alloy casting for greater rigidity, sometimes ribbed for cooling. Internal baffles are fitted to restrict oil surge.

7.2 Liners

The dry liner, a thin sleeve of centrifugally cast iron – usually chill cast or containing chromium for greater hardness – is an interference fit of 0.000 75 mm per mm diameter in the cylinder block. Normally they are produced some 0.4 mm less than the finished bore size and require boring after pressing into place. These dry liners can be used to provide greater wear resistance, to reclaim block castings found to be porous in the bores, or to avoid excessive rebore oversizes after prolonged mileage.

Slip-fit dry liners, finished to size, are sometimes used in original specification and provide for simple replacement. They have a securing top flange and are sometimes copper plated on the outside to improve heat transfer to the block.

The wet liner (Figure 7.1), much thicker and also cast from wear-resisting material, is in direct contact with the coolant and permits a simpler block construction, which may be in cast iron or light alloy,

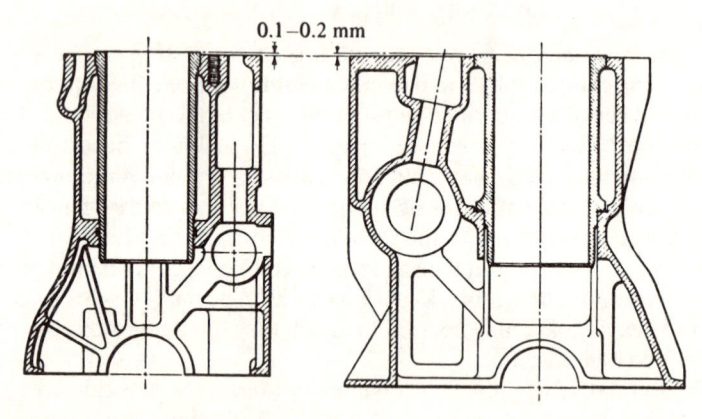

Figure 7.1 Wet-liner construction

together with the advantages of reduced bore wear and easy replacement. Coolant seals are required at top and bottom; the latter are sometimes double O-rings with a drainage channel between them. The upper end of the liner has a projection of some 0.1–0.2 mm above the block face to ensure gas-tight sealing.

In the open-deck arrangement, the top face of the liner – which may be cast into an aluminium block at its base or secured on a lower flange seating – is unsupported by the crankcase mouth and surrounded by coolant. This construction simplifies the block and improves the cooling of the upper liner where it contacts the cylinder head.

Improvements in air and oil filtration, coolant temperature control and piston and piston-ring technology have greatly reduced bore wear in cast-iron cylinders, minimizing the need for liners. They are always required for aluminium-alloy cylinders, except for the limited use of special surface treatments, e.g. direct chromium plating.

7.3 Crankshaft

The stroke/bore ratio for cars is usually about 1:1, although a range from 1.3:1 to the very 'over-square' 0.6:1 has been employed in modern engines.

The crankshaft can be forged from high-tensile steel, or cast from a special alloy iron – nodular or spheroidal graphite – with the advantages of cheapness and reduced machining. Three and four main bearings can be used for four- and six-cylinder engines respectively, or main bearings can be arranged between each crank throw for greater rigidity, reinforced by the overlap of crankpin and journal diameters. A large radius between the journal and the web face increases fatigue resistance and must be retained after regrinding. The journals are often hardened for the later bearing materials. The crank webs may be extended to form counterweights to balance the individual throws and reduce the bearing loads (Figure 7.2).

With long crankshafts the tendency of the power impulses to wind up the shaft and cause torsional vibrations may necessitate a vibration

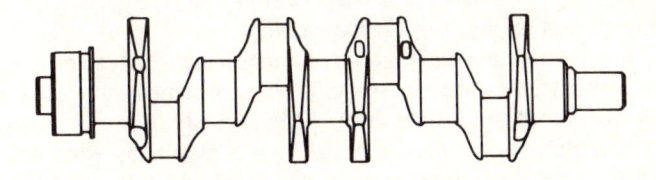

Figure 7.2   Cast-iron crankshaft: overlap on journals

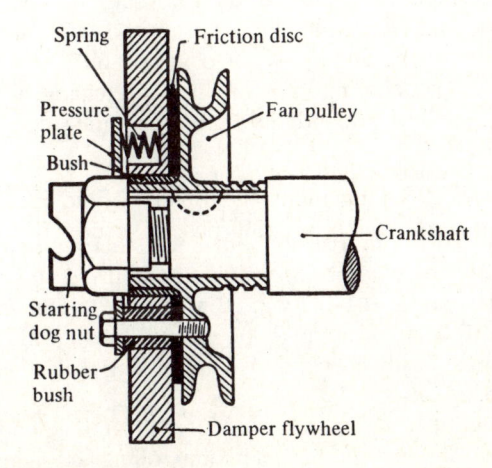

Figure 7.3   Vibration damper

damper. This comprises a small flywheel mounted on the front of the crankshaft and driven through a spring-loaded clutch, bonded or non-bonded rubber or silicone fluid (Figure 7.3). The inertia of the flywheel prevents it responding to the torsional resonance and this energy of the crankshaft is absorbed in friction.

The cast-iron flywheel is usually located on a spigot with dowel pins and secured by bolts to the crankshaft flange. The flywheel provides angular momentum, a carrier for the starter ring gear, and a mounting and friction face for the clutch. Flywheel run-out is usually required to be within 0.000 5 mm per mm of diameter.

7.4 Bearings

The thin-wall bearing consists of a steel backing strip lined with a thin film of bearing metal. Accurate production eliminates hand fitting, gives interchangeability and a high load capacity. The bearing metals are soft relative to the journals and have a low coefficient of friction. Some local flow of the metal can occur to ease high spots or slight misalignment – *conformability*. The surface allows potentially scoring particles of grit or metal to embed themselves – *embeddability*.

As loadings on main and big-end bearings increase, bearing materials of increasing strength are necessary. The aluminium-tin and lead-bronze linings have more than twice the 10 MPa capacity of the original white babbitt's metal. As hardness increases, overlay plating is necessary to provide corrosion resistance and improve embeddability and conformability. Table 7.1 lists various bearing metals.

The bearing shells are located in their housings by a tongue pressed out at one edge and secured by the nip produced when the bearing cap is tightened on to the shells, which protrude 0.025–0.05 mm above the joint face. The bearing caps are non-interchangeable and non-reversible; the shell tongue indents are on the same side of the journal housing (Figure 7.4).

One of the main bearings must also locate the crankshaft axially and take end-thrust from the clutch operation or any helical crankshaft pinion. This usually comprises babbitt-lined steel semi-circular thrust washers fitted to each side of the crankcase housing and bearing cap. The end-float is about 0.1–0.2 mm. Big-end and main bearing clearances are some 0.05 mm – about 0.001 mm per mm of journal diameter. Clearance is most easily checked by compressing a plastigage filament under the bearing cap.

Ball and roller bearings, widely used for main and big-end bearings in single-cylinder construction, have a mainly rolling action instead of the sliding action of a plain bearing (Figures 7.5, 7.6). Consequently the

Table 7.1 Bearing metals in order of increasing hardness and fatigue strength

Material	Percentage composition						
	Tin (Sn)	Lead (Pb)	Copper (Cu)	Antimony (Sb)	Aluminium (Al)	Silicon (S)	
Lead Babbitt	6	84	–	10	–	–	Cheaper than tin Babbitt
Tin Babbitt	90	–	3	7	–	–	Camshaft and thrust bearings
Aluminium-tin	20	–	1	–	79	–	Cheaper than copper-base lining
Aluminium-tin	6	–	1	–	93	–	Overlay plating necessary on these materials; electroplating to 0.025 mm on to the bearing lining
Copper-lead	–	30	70	–	–	–	
Lead-bronze	3	22	75	–	–	–	
Lead-bronze	10	10	80	–	–	–	
Aluminium-silicon	–	–	1	–	88	11	
Overlay	10	84	1	–	–	–	+5 indium

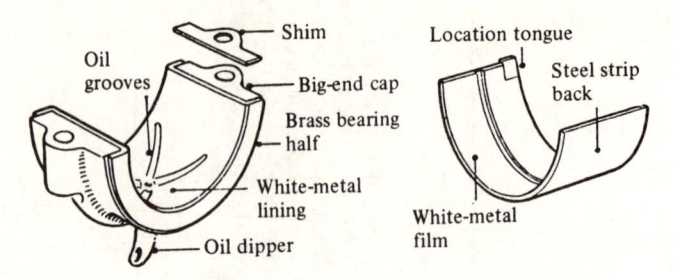

Figure 7.4 White metalled bearings

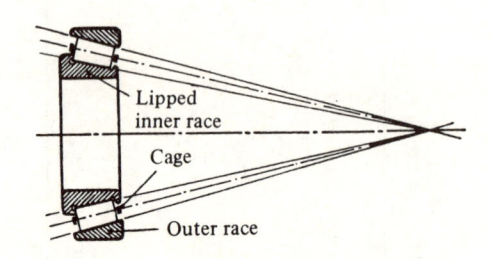

Figure 7.5 Ball and roller bearings

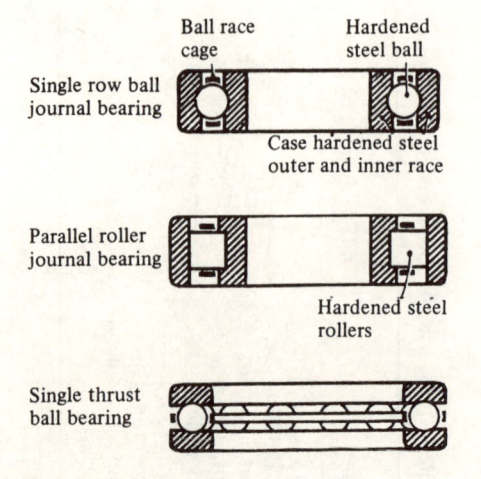

Figure 7.6 Geometry of taper-roller bearing

Engine components

coefficient of friction is lower and comparatively little lubrication is required.

7.5 Connecting rod

The connecting rod carries gas and inertia compressive loading and inertia tensile loads – including the cap fixings. The shank, of I-section for beam strength with lightness, must merge smoothly into the ends to avoid stress concentration (Figure 7.7). A forging in high-tensile steel is general (occasionally light alloy or even titanium), though spheroidal-graphite iron castings are also used. The ratio of the length between centres to the crank throw is between 3:1 and 4:1.

The big end may be split at right angles to the rod centre line or diagonally to permit removal through the cylinder bore. The big-end cap can be located by accurately fitting bolts, dowels or tenons. In the diagonal construction serrations, tenons or hollow dowels are used to relieve the bolts of shear stress.

In the original semi-floating construction the small end has a clamp bolt to secure the gudgeon pin. An interference fit (typically some 0.03 mm) is now usual, and small-end pre-heating (about 250°C) may be used during installation (Figure 7.8).

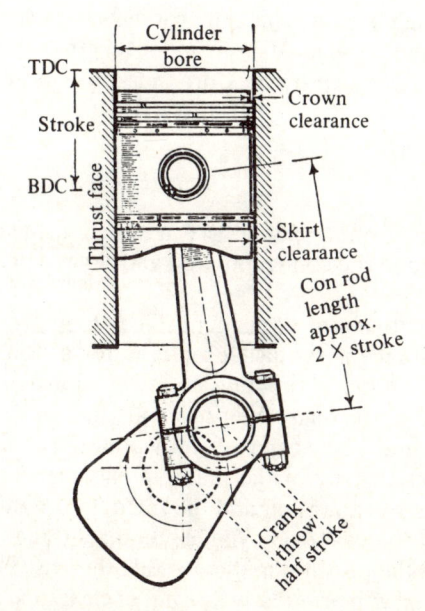

Figure 7.7   Piston and connecting rod

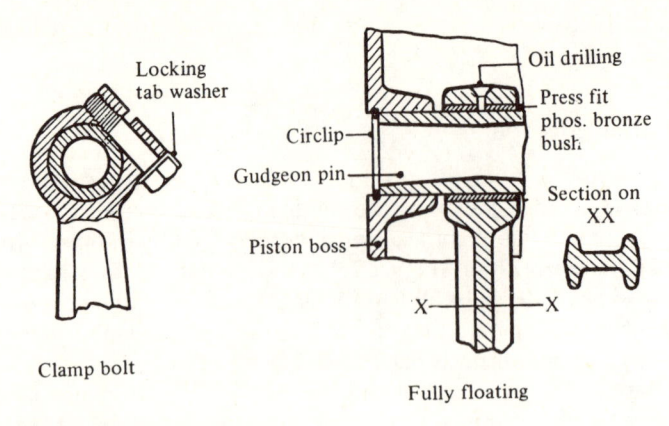

Figure 7.8 Small-end construction

The fully floating arrangement uses an interference-fit steel-backed lead-bronze bush in the small end, with the pin located by piston boss circlips. Gudgeon-pin lubrication is by oil mist, by drain holes from the scraper ring groove or, infrequently, by a pressure feed from the big end which may also incorporate a jet for piston-crown oil cooling.

Connecting-rod alignment can be checked by accurately fitting mandrels through big and small ends or by using a special fixture. Bend or twist should not exceed 0.03 mm in the length of the gudgeon pin. An indication is often given by pressure markings on the side faces of the piston.

7.6 Pistons

The piston must form a sliding, gas-tight seal in the cylinder, transmit the gas pressure to the connecting rod and act as a bearing for the small end.

Aluminium with about 12 per cent silicon to improve tensile strength and wear resistance is the usual material for pistons. It has excellent thermal conductivity, dissipating about two-thirds of the crown heat (some 250–300°C) through the rings and lands and one-third through the skirt; in some cases a jet of lubricating oil is directed under the crown. The low density of the material eases inertia loading, but the high coefficient of expansion and the temperature differential between the piston and the cast-iron cylinder requires design to minimize cold clearances and the piston slap that would otherwise be present.

Taper turning with progressively smaller clearance towards the cooler skirt and oval grinding to give an increased clearance across the

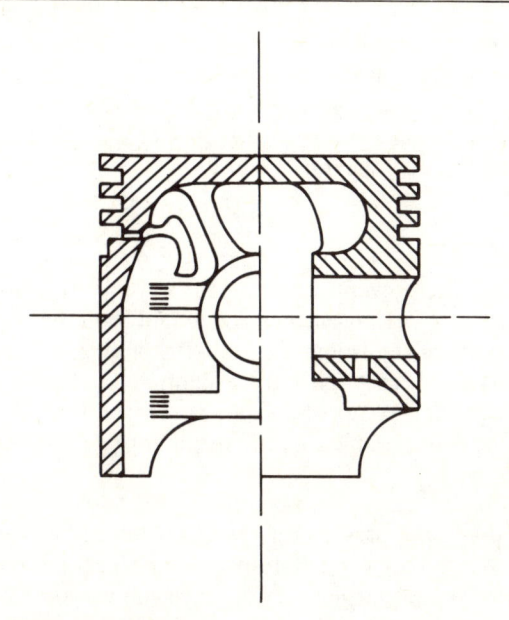

Figure 7.9 Thermal-slot piston

gudgeon-pin axis is normally employed. Heat transfer from the crown can be isolated by thermal slots around the bottom ring groove on both thrust axis faces and diverted, via the strengthening ribs, into the gudgeon-pin boss area (Figure 7.9).

A steel insert, cast into the crown or skirt, can be used to mechanically control expansion. For severe conditions an alloy with 20–22 per cent silicon can be used and will reduce the expansion from 0.000 019 to 0.000 017 per °C. In order to obtain closer limits, selective assembling or grading is often employed. For example, if the manufacturing tolerance on the cylinder bores is 0.030 mm they can be measured and classified into six grades, each having a range of 0.005 mm, and given an identifying letter or number. With a similar procedure applied to the pistons, the assembly of parts bearing the same grade marking will restrict the variation of the piston clearance to 0.010 mm, whereas random assembly of the same parts would permit a maximum variation of 0.060 mm.

In some cases, to reduce noise – by minimizing piston slap – and uneven loading of the skirt thrust face on the power stroke, the gudgeon-pin axis is offset slightly (e.g. 1.5 mm) towards the thrust side of the cylinder.

To prevent abrasion – scuffing – during the running-in period, pistons are often plated with tin or cadmium. Piston clearance may vary from some 0.0005 mm per mm diameter at the open end of the skirt to a possible 0.006 mm per mm diameter for the top land, the clearance being measured at right angles to the gudgeon-pin axis. Ovality on a 75 mm diameter piston can range from about 0.05 mm at the open end of the skirt to 0.2 mm at the top.

7.7 Piston rings

Piston rings serve to prevent gas leakage from, and oil leakage into, the combustion chamber. They account for roughly half the engine friction and half the heat transfer from the piston. Piston rings are almost always of fine-grained special cast iron and are manufactured to exert a uniform radial pressure when compressed into the bore, but the sealing also results from the considerable gas pressure acting behind the ring (Figure 7.10).

The compression rings, whose primary purpose is to prevent gas leakage, have a radial pressure of about 150 kPa. The oil-control ring has a smaller contact area and therefore a much higher pressure – about 700 kPa – and removes surplus oil from the cylinder bore.

A typical petrol-engine arrangement would start with a top compression ring, chromium plated to a depth of about 0.07 mm. This

Standard compression ring may be chromium-plated. Fitted in top groove.

Taper-periphery ring, usually fitted in second groove. Gives quick bedding-in to cylinder bore. May be chromium-plated and fitted in top groove.

Internally-stepped ring, causing a twist and taper periphery when compressed and fitted in second groove.

Externally-stepped ring for compression and additional oil control. Fitted in second groove.

Slotted standard oil-control ring.

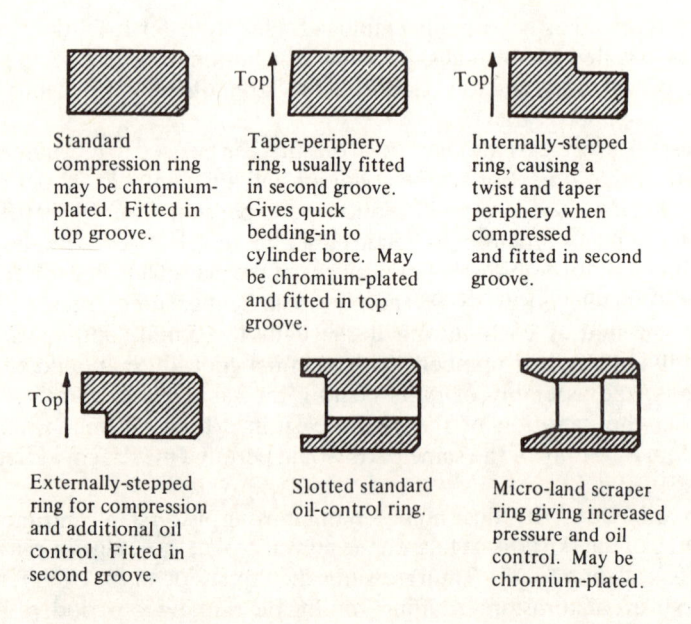

Micro-land scraper ring giving increased pressure and oil control. May be chromium-plated.

Figure 7.10   Piston rings

reduces bore wear by at least half, as the surface is too hard to pick up abrasive particles and, by a burnishing action, improves the cylinder wall surface. A molybdenum-filled inlay is also used for the same purpose. The second ring may have an upward-facing taper to promote rapid bedding-in and oil scraping; the taper is often produced by a step in the internal upper surface causing the ring to twist when fitted. A slotted oil-control ring fitted above the gudgeon pin completes the pack. Alternatively the oil-control ring may be a composite of two or more flexible steel rails with chromium-plated rounded edges spaced and forced against the cylinder wall by a spring expander.

The piston rings may be phosphate treated or, if chromium plated, coated with a running-in compound to ensure the important bedding-in process during the first 1000 km. This is often promoted by plateau honing of the cylinder bore to leave a criss-cross of fine oil-retaining scratches on the load-bearing plateau.

In replacement ring packs the top ring may have a step in the top edge to avoid contact with any wear ridge at the upper limit of ring travel.

Diesel pistons can use an additional compression ring, and for heavy duty the rings may be carried in a bonded-in wear-resistant alloy cast-iron carrier.

For water-cooled engines a ring gap of about 0.003 mm per mm diameter is required to allow for expansion. The groove depth should be about 0.25 mm greater than the radial thickness of the ring and a side clearance of some 0.05 mm is necessary.

7.8 Camshaft

The camshaft needs to be stiff and well supported since it is subject to heavy shock loads. The force required to accelerate a valve with a mass of 100 g may be 1–1.5 kN at maximum engine revolutions.

An alloy cast iron is usual for the camshaft. The insertion of chills into the mould enables hardened cam and bearing surfaces to be produced. Bearings may be placed between each pair of cams or between each set of four. Lined steel shell bearings can be used, or the camshaft can run directly in the iron or aluminium-alloy casting.

An overhead camshaft (OHC) location reduces flexure and some inertia in the valve train and simplifies the block casting at the cost of extending the drive and lubrication arrangements. The camshaft drive may be through gears – normally helical, chain, toothed belt or some combination of these (Figures 7.11, 7.12). Where required a manual fixed adjustment, or a spring-loaded and sometimes oil-assisted ratchet automatic tensioner, is provided on the non-driving side of the chain or belt.

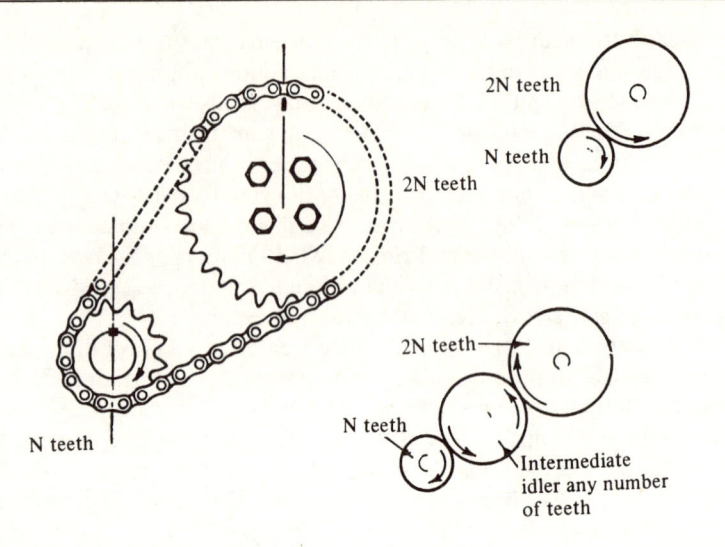

Figure 7.11   Camshaft drive

The toothed belt, of neoprene over a high-tensile steel or glass-fibre core – and requiring no lubrication – has promoted the popularity of OHC for car engines. An adequate piston-to-valve clearance should allow belt failure without valve damage. The belt cost is modest, and replacement at some 50 000 km is a useful safeguard.

The cam profiles are designed to provide, by the rate, height and duration of valve lift, the optimum volumetric efficiency for the speed range of the engine, whilst maintaining acceptable accelerations in the valve train.

The overhead camshaft can act directly upon the valves through chilled alloy cast-iron inverted bucket tappets or by rockers. The overhead-valve (OHV) train employs tappets, push rods and rockers. In some cases the tappets are arranged to rotate under cam action to distribute wear (Figures 7.13, 7.14, 7.15).

Rockers can pivot on a longitudinal shaft supported by pedestals, or on individual ball pivots. The rockers can be of malleable cast iron or forged steel, or fabricated from steel pressings.

For valve clearance adjustment the inverted bucket tappets can have interchangeable thick shims beneath the head or, for greater accessibility, in the top recess. An internal screwed adjustment bearing on to the end of the valve stem is another convenient alternative. Rockers may have a screwed adjustment, an eccentric bush or an adjustable ball-pivot height.

| *Engine components*

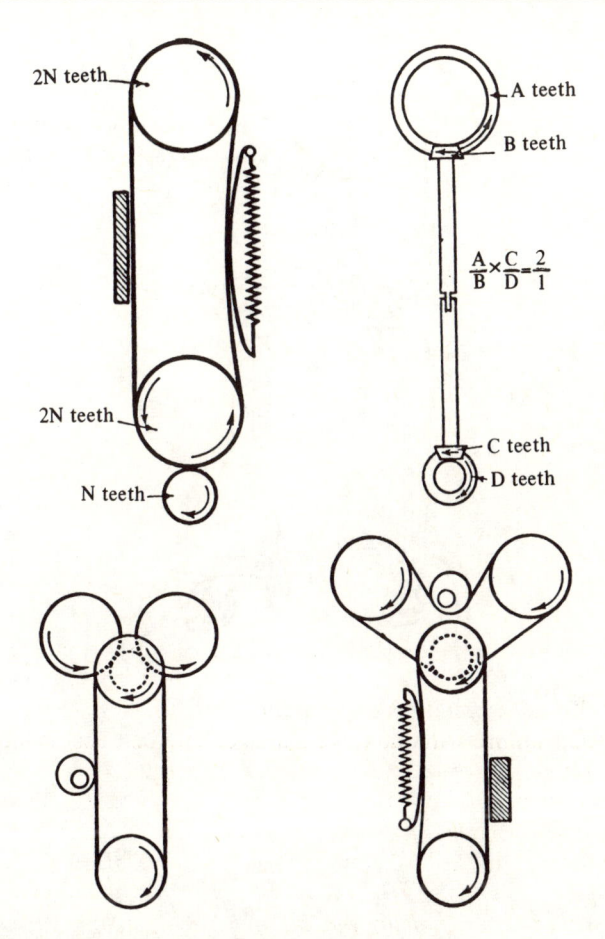

Figure 7.12   Overhead-camshaft drive arrangements

Valve clearance must be set when the cam lobe is opposite the tappet or rocker. This is readily observed with OHC.

Pistons moving in unison in an engine are 360° out of phase and their cams are 180° out of phase, i.e. opposite. When setting OHV clearances it is therefore only necessary to ensure that the corresponding valve in a cylinder with the piston moving in unison is fully open. However, the valves must be correctly identified, inlet with inlet and exhaust with exhaust. For example, on a four-cylinder in-line engine, number 1 valve can be adjusted when number 7 or number 8 is fully open, depending upon the port arrangement.

101

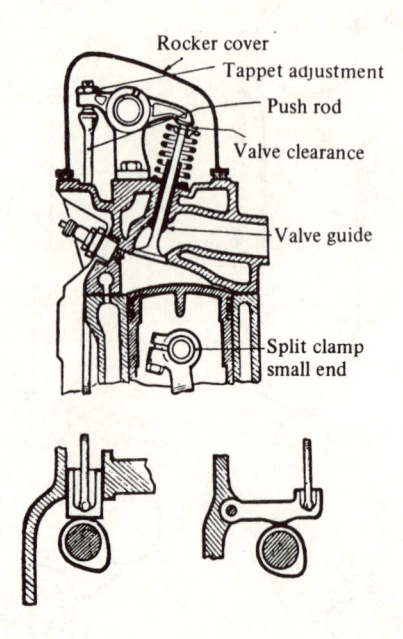

Figure 7.13 Overhead valve operation

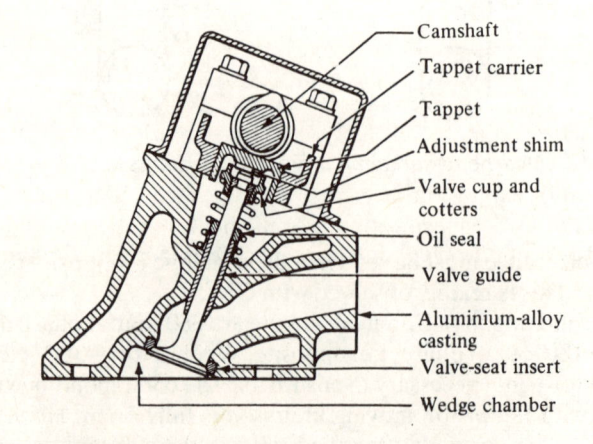

Figure 7.14 Overhead camshaft: inverted bucket tappet

Engine components

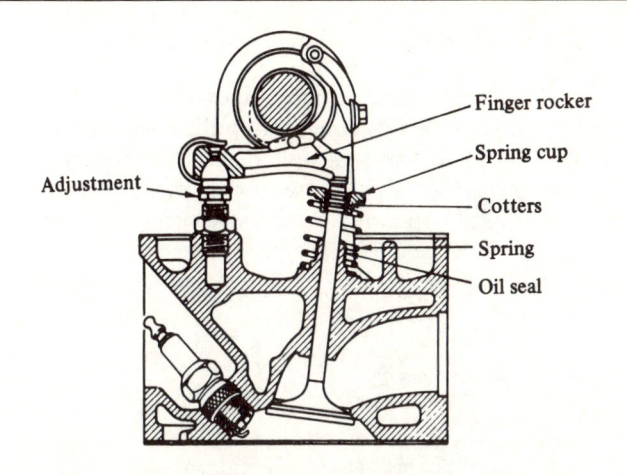

Figure 7.15   Overhead camshaft: pivoted follower

Some makers do not use extended quietening ramps on the cam profile and both valve clearances can be set with the piston TDC compression stroke.

7.9   Valves

Valves operate under very arduous conditions, as the gas temperature may exceed 2000°C. The exhaust valve head is normally the hottest part of the combustion chamber, with a possible temperature over 800°C.

Forged silchrome steels are satisfactory for inlet valve temperatures up to about 650°C, and contain chromium with silicon and nickel. A 21-12 or 21-4 chromium-nickel stainless austenitic steel may be used for exhaust valves operating up to some 800°C. Because of cost and a high coefficient of expansion the valve-head forging is often friction welded to a stem of lower specification. For the highest duties nimonic alloys – some 70 per cent nickel and 20 per cent chromium, with titanium, cobalt and manganese – may be used.

In some cases protective coatings enable less costly materials to be used, such as an aluminium overlay for inlet valves, and stellite (60 per cent cobalt, 29 per cent chromium with tungsten) for exhaust seatings.

The usual valve seat angle is 45°, but occasionally 30° for inlet only. The seat width is a compromise between sealing pressure and conductivity. Alloy cast-iron valve-seat inserts are often needed as a repair measure. Inserts are essential with an aluminium-alloy cylinder head, and an austenitic cast iron is required with a high coefficient of expansion. Inserts are usually an interference fit using cryogenic techniques (e.g. shrinking in liquid oxygen).

The valve stems may operate either in cast-iron guides pressed into the cylinder head or directly in the head. If the guides are formed in the head, reaming and oversize valves are used to rectify wear. After reaming or when new valve guides have been fitted, the valve seats must be recut. Oil seals incorporated in the valve spring cup, secured above the valve guide, or of the umbrella type mounted on the valve stem, are usually essential to prevent oil entering the combustion chamber.

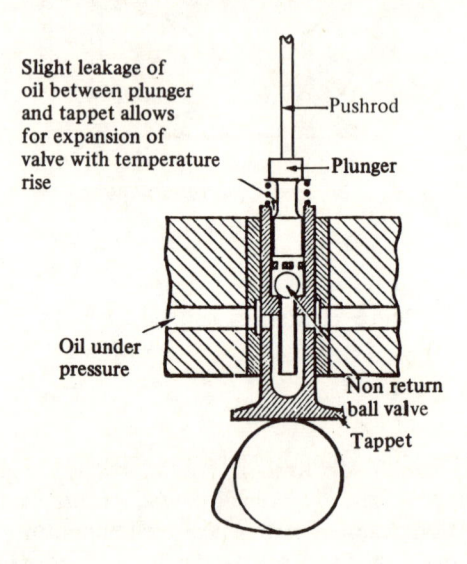

Figure 7.16   Hydraulic self-adjusting tappet

Valve clearance is necessary to allow the valves to seat under all conditions; it varies with engine design. Cold clearance averages about 0.25 mm – the exhaust having the greater clearance where a difference occurs. Insufficient clearance may cause a loss of compression, some spitting-back through the inlet valve, or a burned exhaust valve. Excessive clearances result in insufficient lift and can impose excessive loads on the valve gear.

Hydraulic self-adjusting tappets have a light spring-loaded extension taking out all clearance from the valve train. Trapped oil maintains the extension during valve lift but a designed leakage allows for valve expansion. Noise and adjustment are eliminated and maximum lift is always available (Figure 7.16).

7.10 Valve springs

The valve springs must return the valves and the operating gear and maintain the tappets or rockers in contact with the cams at the highest engine speed employed. If not the valves will close late, reducing power owing to the lowered volumetric efficiency. There is also the danger on some engines of the piston striking the valves – with certain damage.

If the natural frequency of a valve spring should respond to the valve movement, a surge may travel through the coils, delaying high-speed closure. This response can be limited by using a progressive pitch on the

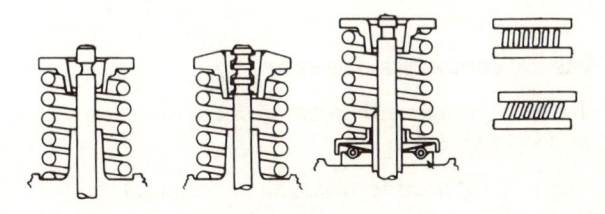

Figure 7.17   Valve-spring retention: non-rotative; top rotation; base rotation

coils. The closer coils are fitted nearest to the cylinder head. Double springs have the disadvantage of lower individual rates and lower natural frequency, and consequently greater likelihood of resonant surge. However, they offer protection against one spring breaking, and some mutual damping can be provided between their coils.

Valve-spring material is either high-carbon or alloy steel, mechanically treated by pre-stressing in torsion, scragging and then shot blasting to increase fatigue strength – as with suspension coil springs.

The method of valve-spring retention depends upon whether valve rotation – desirable to even heat distribution and clear seat deposits – is to be achieved at the top or the bottom of the valve spring (Figure 7.17). If the former, the abutting collets in the valve-spring retainer will allow a small radial clearance on the valve stem so that vibration can produce a gradual rotation when lifted. If rotation is achieved by a rotator beneath the valve spring – a garter-spring or ball-ramp device – then the tapered collets wedge together the valve stem and the valve-spring retainer.

7.11   Joints

Gaskets may spread the sealing pressure evenly over the mating surfaces or, by restricting the contact area, concentrate a higher pressure on a sealing ring. For the cylinder head–block joint the copper-asbestos combination has been superseded by the cheaper corrugated-steel, laminated-steel or various composite asbestos-rubber-steel core gaskets.

In all cases the mating surfaces must be tested for flatness and checked for imperfections before assembly.

Tightening techniques may involve specification of the thread lubrication, the tightening sequence, the progressive and final torques or a prescribed angular rotation after a maximum torque, with any subsequent retightening. In general, tightening involves repeatedly working outwards until the required torque is achieved.

7.12 Testing compression pressure

The procedure for using the compression gauge to assess cylinder condition is as follows:

1. The engine must be at normal running temperature.
2. All sparking plugs are removed.
3. Throttle valve is secured in fully open position.
4. Insert the compression gauge in each sparking plug hole in turn.
5. Rotate the engine on the starter motor, which must revolve at normal speed. Record for each cylinder:
   (a) Pressure recorded on first compression stroke
   (b) Number of compression strokes required to obtain maximum pressure
   (c) Maximum pressure.
6. If necessary, inject a small quantity of oil above the piston and repeat test 5. The addition of the oil will improve the readings of a cylinder by some 35–70 kPa by sealing the piston rings.
7. Maximum readings will be lower than the manufacturer's figures by some 34 kPa for each 300 m above sea level.

Assessment details are given in Table 7.2.

7.13 The vacuum gauge

The vacuum gauge connected to the induction manifold can indicate various engine conditions, including faulty piston ring or valve sealing and air leaks into the manifold system.

The engine must be at the normal operating temperature before testing. For altitudes over 600 m above sea level the readings will show a reduction of some 3.3 kPa for each 300 m increase in height.

Vacuum–gauge assessment details are given in Table 7.3 (page 106).

Table 7.2 Compression-gauge cylinder assessment

First reading	Build-up of pressure	Maximum pressure	Pressure variation between cylinders	Indications
Normal	Prompt	Within maker's limits	Within 10%	Correct. Some makers specify pressure variation between cylinders to be within 5%
Low	Fair	Slightly below maker's limits	Uniformly low on *all* cylinders	Worn rings or bores. Confirmed if small injection of oil considerably improves first reading
			On *one* cylinder	Broken rings. Confirm as above
Low	Slow	Much below maker's limits	One or more cylinders	Valves or seats defective. Confirmed if small injection of oil does *not* improve readings very much
			Two *adjacent* cylinders give same readings	Cylinder-head gasket defective between cylinders

Table 7.3 Vacuum-gauge engine assessment

Engine (rev/min)	Gauge reading (kPa below atmos.)	Indication
Cranked by starter motor. Throttle valve fully closed	40–60, steady	Correct
	Below 40, steady or falling	Air leaks in induction system; flanges, gaskets, valve guides, throttle spindle faulty
	Below 40, unsteady	Valves faulty. Test compression
Idling	60–68, steady	Correct. At very low idling speed some fluctuation is normal
	Low, unsteady	Incorrect mixture. Adjust to obtain highest vacuum. Check fuel level, pump pressure
	Intermittent fall of 10–14	One valve faulty. Test compression
Fast idle. Short each plug in turn	Equal falls	Correct
	Unequal	Valves, rings or ignition defect. Faulty cylinder produces least fall
Accelerate	Falls 16 or more then rises	Correct
	Fall less than 16	Air cleaner choked. Remove and retest
Accelerate to half speed and close throttle	Fall followed by prompt rise to normal or above	Correct
	Sluggish action with small fluctuations	Choked exhaust system. Check outlet for pulsations of gas flow
Accelerate to maximum speed and close throttle	Fall to near zero and prompt rise to 10–20 above normal	Correct
	Sluggish action	Worn rings and bores
Slow acceleration to maximum speed	Steady	Correct
	Fluctuations of 34–78 increasing with speed	Weak or broken valve springs
Constant speed 2000–3000	60–80, steady	Correct

Table 7.3

Engine (rev/min)	Gauge reading (kPa below atmos.)	Indication
	Low, steady	Retarded. Test with timing lamp. Poor compression. Test. Incorrect valve timing. Check valve clearances and timing
	Fluctuating	Electrical defects, plugs, contact breaker, coil or condenser
	Falls slowly	Choked exhaust system

Chapter 8
The lubrication system

8.1 Lubrication

Lubrication is necessary to minimize friction and wear between moving surfaces and eliminate the risk of seizure. Fluid lubrication is the ideal condition where the bearing surfaces are completely separated by a layer of oil, the resistance to movement depending upon the viscosity of the lubricant. *Oiliness* is the tendency of the oil molecules to adhere and bond to a metallic surface. This bonding of oil to a moving surface causes an attached layer to be dragged along. When a journal revolves in a bearing these moving layers approach the area of minimum or zero clearance and build up a hydrodynamic oil wedge separating the surfaces and floating the shaft on a substantial film of oil.

Greasy or *boundary* lubrication occurs where the oil film is very thin and is insufficient to completely separate the two surfaces, such as piston and cylinder or valve stem and guide. Under these severe conditions good oiliness and film strength is essential. An oil with a vegetable base, e.g. castor oil, can be superior to mineral oil in these respects, but has other disadvantages and is seldom used.

Dry friction takes place in the absence of any lubricant and would quickly produce overheating and failure of a loaded bearing.

8.2 Viscosity

The resistance to flow or viscosity of an oil decreases with a rise in temperature. It is measured by the SAE (Society of Automotive Engineers) number, which is related to the number of seconds required for a sample to flow through an orifice under standard conditions. SAE 20, 30, 40 and 50 refer to 210°F (98.89°C) at the highest desirable operating conditions. SAE 5W, 10W and 20W refer to 0°F (-17.78°C) for cold starting performance. The SAE number relates solely to viscosity and has no reference to other properties of the oil.

The lubrication system

8.3 Viscosity index

The rate of change of viscosity with temperature is measured by the viscosity index. An oil with a high viscosity index, that is one having a relatively small reduction of viscosity with rise in temperature, is desirable. Easy starting from cold and rapid distribution of oil throughout the engine require a low viscosity, but the viscosity must be reasonably maintained at running temperatures otherwise the oil film may fail or the oil consumption become excessive.

8.4 Additives

The viscosity index (VI) of an oil can be raised by the addition of viscous resins. In this way it can combine the viscosity of, say, an SAE 20W oil at 0°F with that of an SAE 50 oil at 210°F, and would then be given the multigrading SAE 20W/50. The greater the range of numbers in the multigrading, the higher the VI of the oil.

A detergent-dispersant cleansing action, minimizing the formation of sludge or carbon deposits, is obtained by the addition of calcium, barium or other metallic soaps. The contaminants, which would otherwise deposit on the internal surfaces of the engine, are held in suspension in the oil and are removed by filtration or during the oil change.

Oil oxidation is inhibited by the use of phosphorus or zinc compounds. Substances such as colloidal graphite, zinc oxide and molybdenum disulphide are valuable additives in maintaining a lubricating film under boundary conditions.

8.5 Splash lubrication

This early arrangement is now retained for small industrial engines where the small four-stroke unit, often in side-valve form, offers an economic alternative to a two-stroke engine.

The big ends carry dippers which enter trays of oil feeding the big-end bearings. Oil is splashed upwards to lubricate the cylinder walls and, collecting in galleries, feeds to the main and camshaft bearings and to the timing gear. The trays are filled from oil spray collected from the revolving flywheel (Figure 8.1).

8.6 Force-feed lubrication

A typical car system has a sump of some 3.5 litres capacity and employs by-pass (Figure 8.2) or full-flow (Figure 8.3) filtration. Pressure is

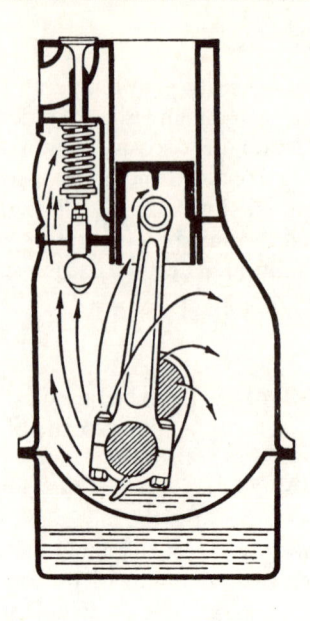

Figure 8.1 Splash lubrication

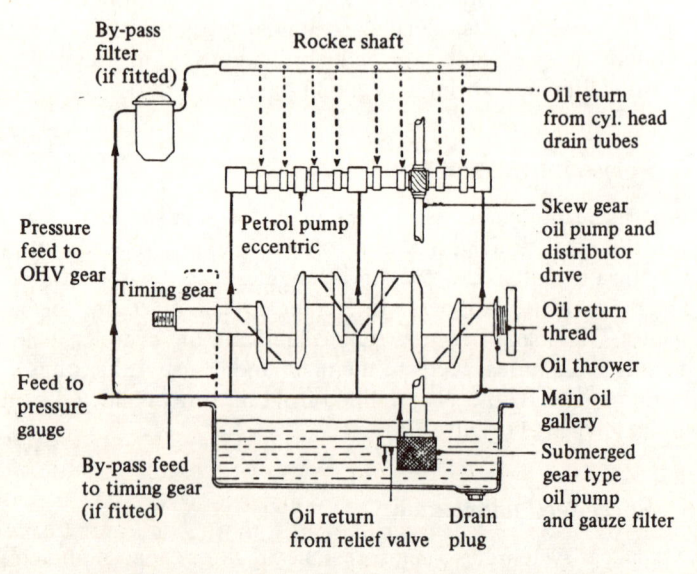

Figure 8.2 Pressure lubrication: by-pass filtration (schematic)

The lubrication system

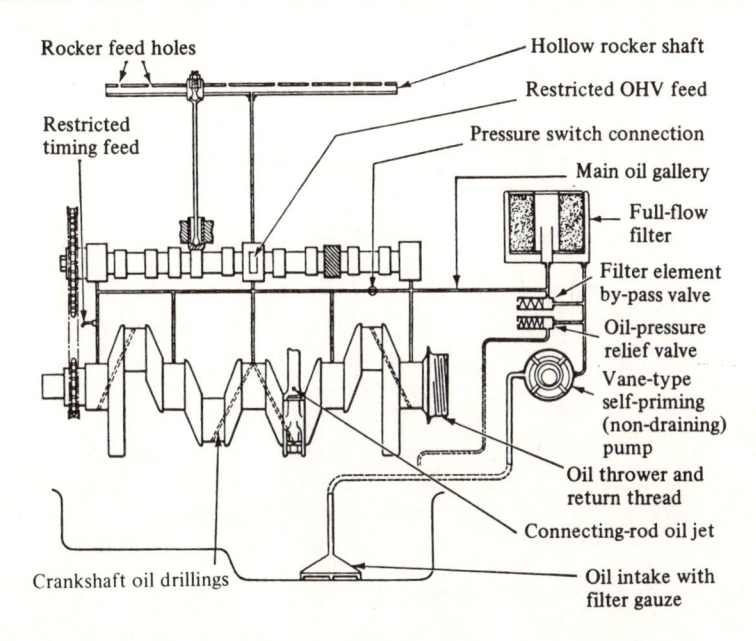

Figure 8.3 Pressure lubrication: full-flow filtration (schematic)

maintained by an eccentric-lobe, external-gear, internal-gear crescent or sliding-vane pump (Figures 8.4, 8.5, 8.6). The pump may be driven by skew gearing from the camshaft (Figure 8.7) or crankshaft or directly off the end of the camshaft; the crescent-type pump is mounted on the crankshaft ahead of the front main bearing. If a high mounting is adopted, a weir and suitable port arrangement ensures that the pump retains oil and is self-priming.

The pump intake is protected by a gauze strainer and the oil is delivered through a full-flow filter to the main oil gallery, a drilling in the crankcase with branches to each main and camshaft bearing. Oil pressure is created by the resistance of the system – the bearing clearances and metered supplies – and is limited by a relief valve to some 300–400 kPa. The big-end bearings are pressure fed through drillings in the crankshaft communicating with a central groove around the main bearing shells. From the big-end bearings oil sprays on to the cylinder walls, pistons, small ends and cams and creates a mist, assisting lubrication of the cams, tappets and camshaft drive.

The overhead valve gear is lubricated through a hollow rocker shaft receiving a reduced supply, often by means of a flat or drilling in one of the camshaft journals which pressurizes the feed once per revolution.

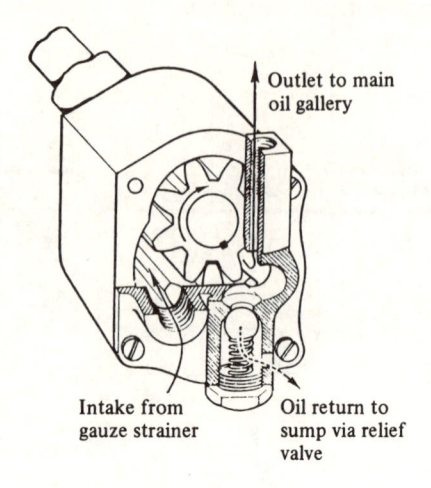

Figure 8.4   External-gear oil pump

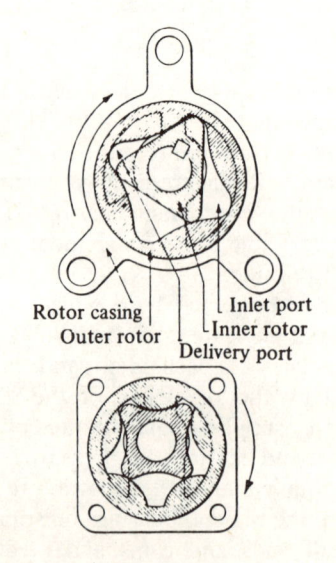

Figure 8.5   Eccentric-lobe oil pump

The lubrication system

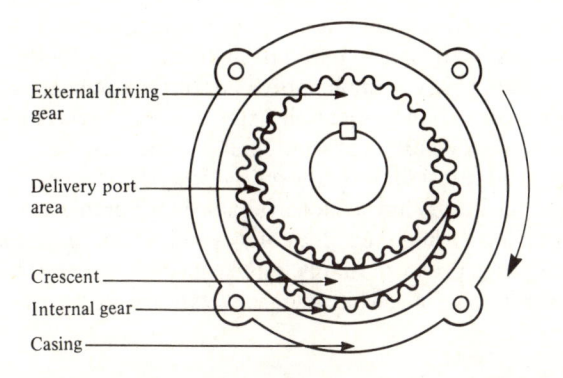

Figure 8.6 Internal-gear crescent oil pump

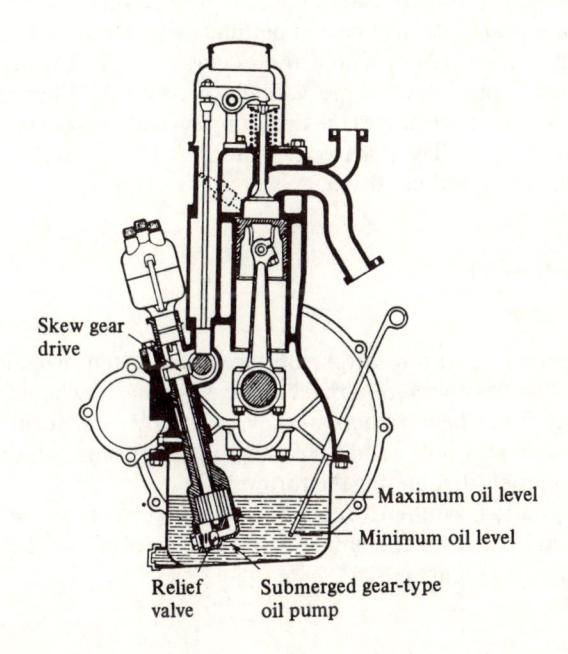

Figure 8.7 Pressure lubrication system

Drillings in each rocker provide lubrication for the push-rod cups and valve stems, the oil draining back and lubricating the tappets and cams. Where ball-pivot rockers are used these can be lubricated from hollow mounting studs fed from a longitudinal gallery.

A metered supply is normally provided for the camshaft drive chain. Some engines also employ a jet hole in the connecting rod for additional lubrication of the cylinder walls and small-end bearings.

The overhead camshaft requires special consideration because it does not have the subsidiary lubrication available in the crankcase. Typically, a vertical drilling through the block and cylinder head from the main oil gallery provides a pressure supply to the camshaft bearings via drillings in the camshaft pedestals, while metered jets lubricate the cam surfaces.

8.7 Dry-sump lubrication

The dry-sump system employs a scavenge oil pump, of larger capacity than the supply pump, to return possibly aerated oil from the crankcase to a separate oil tank. The system permits a smaller crankcase, reduces oil splash and oxidation within the engine and, with a suitable tank arrangement, provides a large volume of cool oil. Where an engine operates at abnormal angles the dry-sump system prevents interruption of the oil supply. Twin scavenge pumps fed from each end of the crankcase are sometimes fitted.

8.8 Oil sealing

Dynamic seals

An oil thrower and a return scroll thread are often machined on the crankshaft between the rear main bearing and the flywheel flange. The oil is flung from the periphery of the thrower by centrifugal force, and the thread, having only a small clearance in its housing, returns the oil to a drain channel surrounding the thrower.

Spring-loaded synthetic rubber seals, in conjunction with an oil thrower, are usual for the timing cover and are also used at the rear of the crankshaft (Figure 8.8).

Static seals

Static seals may include cork, paper or synthetic gasket material used either on plain surfaces or on surfaces relieved to increase the sealing pressure. Additional sealer compound is sometimes required, e.g. in the cap area of the rear main bearing on some designs.

The lubrication system

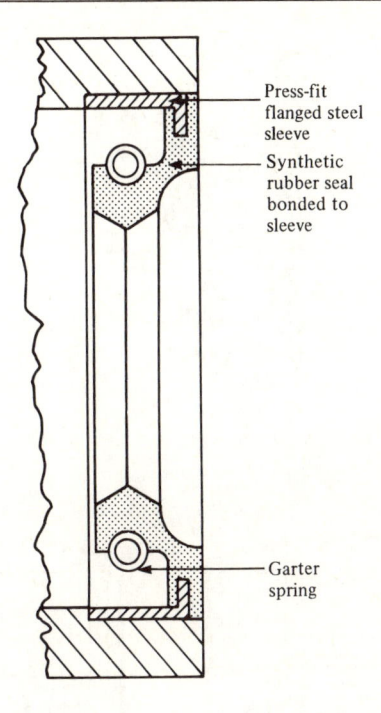

Figure 8.8 Spring-loaded lip seal

8.9 Oil filtration

On the inlet side of the oil pump the pressure difference is small since it depends upon atmospheric pressure and only a surface filter can be fitted – a gauze strainer.

On the outlet side of the pump the maximum pressure is determined by the relief valve setting. Without a relief valve dangerously high pressure can be produced, particularly with cold oil, which can shear the pump drive or burst filter canisters (Figure 8.9).

Current practice is to filter all the oil leaving the pump by a full-flow depth filter; the particles are retained in the depth of the filter material. Flow resistance and fine filtration must be balanced with a full-flow system; resin-impregnated paper can remove particles down to 5 μm. The usual form is a paper element, pleated into a multipoint star to increase the filtration area, through which the oil passes from outside to inside. A relief valve is incorporated, set at 70–100 kPa to allow circulation should the element become choked (Figure 8.10).

117

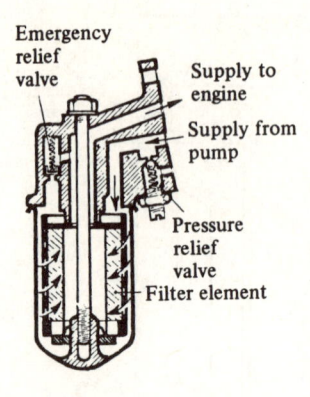

Figure 8.9 Full-flow filter: separate element and relief valve

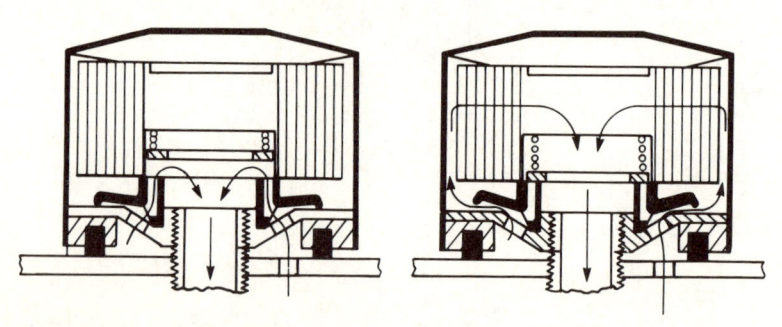

Figure 8.10 Full-flow canister filter

By-pass filter

Since a by-pass filter deals with only a part of the oil circulating, total filtration – if attainable – takes much longer. It is not suitable where abrasive particles must be excluded from the harder bearing materials of reduced embeddability.

Separation

In some engines centrifugal separation has been used to deposit contaminating particles in the periphery of a high-speed rotating chamber, e.g. a modified crankshaft front pulley.

A permanent magnet, often incorporated in the drain plug, can remove ferrous particles from the circulation of oil.

The lubrication system

8.10 Oil pressure indication

A crankcase-mounted pressure switch controls a fascia panel warning lamp. When the pressure in the main oil gallery reaches some 50–70 kPa a diaphragm in the switch opens the contacts (Figure 8.11). The oil warning light gives no indication of the quantity of oil in the engine – it may operate satisfactorily with the sump almost empty – nor does it necessarily indicate that oil is reaching all the pressure-fed bearings.

In some cases, a pressure-differential switch is fitted in the full-flow filter to indicate when the element is becoming choked; this switch may control the normal oil warning light or a separate indicator.

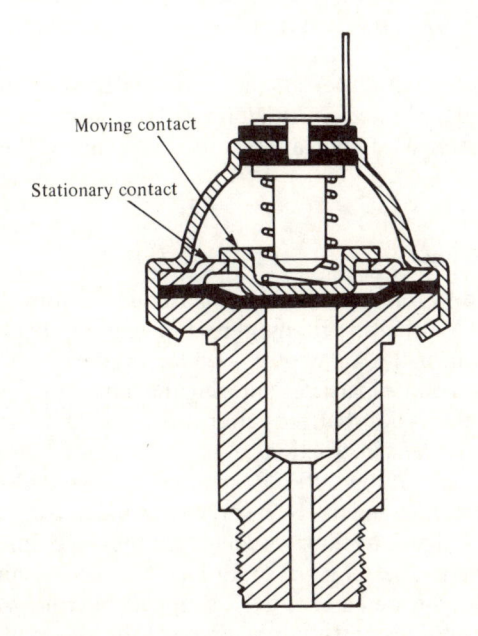

Figure 8.11 Oil pressure switch

8.11 Oil cooling

The entire contents of the sump are circulated many times a minute; the pump delivers 10–20 l/min at 1000 rev/min. This has an important engine cooling effect as well as lubricating the bearing surfaces. In some cases the oil is specifically utilized for heat transfer, as for example by directing a jet at the underside of the piston crown.

To maintain a satisfactory oil temperature – below 100°C – the main oil flow may be directed through an oil radiator situated in the cooling air. This is standard practice on many air-cooled vehicle engines. Some heavy vehicles use an oil/cooling-water heat exchanger.

8.12 Oil dilution and sludge

Short runs and repeated cold starting promote oil dilution from unvaporized fuel in the cylinders.

Sludge is produced by oxidization and contamination of the oil with carbon, dust, metallic particles and condensed water and acid vapour; approximately 1 litre of water is produced (as steam) from the burning of 1 litre of fuel. The excessive blow-by of gas past piston rings in a worn engine (normally some 0.025 m^3 per minute) encourages rapid sludge formation.

The use of detergent-dispersant lubricant (adequately filtered and regularly changed), crankcase ventilation and the prompt attainment of running temperature are important factors in minimizing the formation of sludge.

8.13 Oil changing

Newly machined surfaces, although apparently smooth, are microscopically rough. During the running-in of new or overhauled engines the bedding-down of the high spots – under progressively increasing loads – results in metallic particles entering the lubricating system. The oil should therefore be changed after the first 1500 or 3000 km, according to the maker's instructions.

Normal oil changes at some 10 000 km intervals should be carried out when the oil is hot, thin and likely to carry away the maximum quantity of accumulated sludge. The filter is renewed at the same time.

Where required, the engine condition can be monitored and abnormal functioning detected – e.g. incipient bearing failure – by regular chemical analysis of the lubricating oil.

Blue smoke when accelerating after a period of engine over-run or idling indicates oil entering the cylinders from worn cylinder bores and rings or defective valve oil seals. Oil consumption is related to viscosity, and consumption may rise with prolonged high speeds and associated temperature.

8.14 Grease

Grease is normally a mineral oil thickened with a metallic soap plus additives to resist oxidation, corrosion, scuffing, high temperature or extreme pressure.

The lubrication system

Calcium grease (water resistant and with a 50°C temperature limit) and sodium grease (water soluble and with a 100°C temperature limit), though still in use, have largely been replaced by lithium-base grease, which combines water resistance with a high temperature limit of about 120°C. Molybdenum sulphide, graphite, zinc oxide or copper, all in a finely divided form, may be added to provide specific resistances.

Other bases can be used, including aluminium, strontium or barium metallic soaps, whilst a silica- or clay-base grease can provide lubrication up to 250°C.

When using a pressure grease-gun, consideration must be given to the possibility of over-charging and seal damage. Ball or roller bearings must never be fully packed with grease, which will cause overheating and possible damage.

Chapter 9
The cooling system

9.1 Heat dissipation

About one-third of the heat energy of the fuel must be dissipated in the cooling system to prevent overheating and eventual failure of the valves or piston. Less severe overheating may reduce volumetric efficiency owing to a lower charge density, and may cause detonation, pre-ignition and lubrication difficulties.

Since excessive cooling also results in a lower thermal efficiency with fuel vaporization and oil contamination problems, there is an optimum running temperature, and as the ambient air temperature varies considerably the degree of cooling should be controllable.

All engines are directly or indirectly air cooled, but in most cases the engine heat is transferred by a liquid coolant.

9.2 Air cooling

Finned cylinders, often of aluminium alloy, linered or with a special chromium-plated or nickel-carbide surface, are widely used on motorcycles and small industrial engines (Figure 9.1). Machine movement, or a flywheel fan and shroud, provides adequate air flow for these applications. For multi-cylinder vehicle engines a separately driven centrifugal or axial flow fan and adequate cowling are needed. This will absorb substantial power at high speed. An oil radiator and temperature control of air flow may be required. An efficient interior heating system is complicated and, without the blanketing effect of coolant jackets, mechanical noise will increase. The system offers freedom from coolant problems, the opportunity of a higher running temperature, reached quickly from cold, and possibly a slight weight reduction.

9.3 Thermosyphon water cooling

Heat from the combustion of the charge is conducted through the walls of the cylinder and head to water in cored passages surrounding these

The cooling system

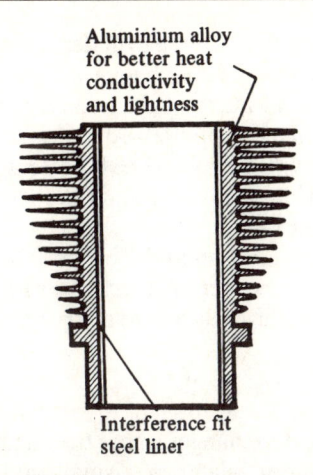

Figure 9.1 Linered air-cooled cylinder

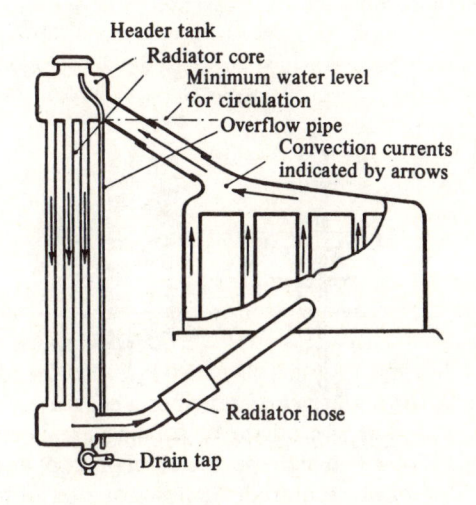

Figure 9.2 Thermosyphon cooling system

parts. The cooling water, becoming less dense as it is heated, moves in a convection current towards the radiator header tank, being displaced by colder and denser water from the radiator base tank. Passing downwards through the radiator core, the heat is transferred to the air current created by the movement of the vehicle and assisted by a fan. This circulation is known as thermosyphon cooling (Figure 9.2).

The heat conducted through the cylinder walls is more readily transferred to the water than to air and, having a high specific heat capacity, a relatively small mass of water will absorb and dissipate a large quantity of heat. When the heat dissipated from the radiator by conduction, convection and radiation equals the heat entering from the cylinders, a condition of equilibrium is obtained and the temperatures around the circuit will remain constant.

With the thermosyphon system the cylinder head should be situated well below the radiator header tank to assist convection, and the water level must never fall sufficiently low to interrupt the circulation.

9.4 Radiators

The radiator is the heat exchanger where the engine heat is dissipated to the air, with the interior heater as a small subsidiary. The metal/air interface should greatly exceed that of the water/metal area. Separate water tubes were fitted with plain or corrugated fins or wire windings for this purpose (Figure 9.3).

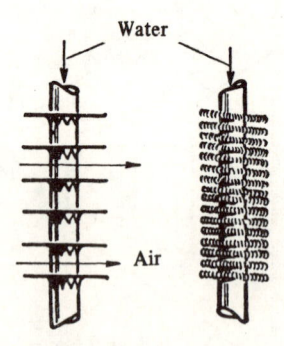

Figure 9.3   Radiator tubes: finned, wire wound

The tubular-type matrix usually employs folded and sweated (soldered) thin brass strip to form the tubes, which are flattened to increase the surface/volume ratio of the water. These can pierce a stack of thin copper air fins, which may be dimpled or louvred to improve air contact. Alternatively the water tubes may have deeply corrugated air fins sweated between them. The film-type core has a high heat-dissipation/weight ratio but a lower resistance to internal pressure and is less suitable for pressurized systems (Figure 9.4).

The cooling system

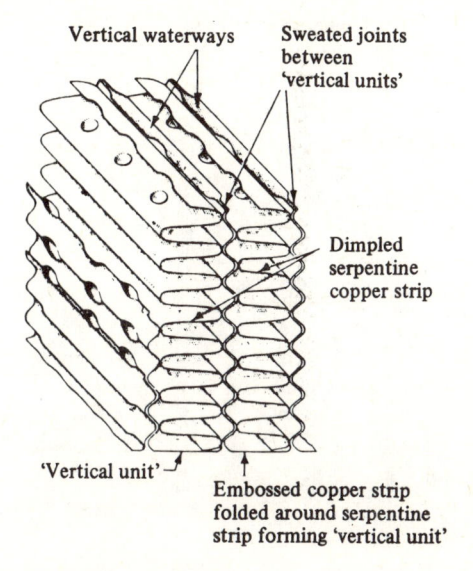

Figure 9.4   Film-type radiator core

9.5   Water pump

A centrifugal pump or impeller, belt-driven from the crankshaft, is fitted to the cylinder block or head to increase the rate of coolant circulation (Figure 9.5). When a water pump is used a smaller radiator can be fitted, together with an interior heater and thermostat. The system can be pressurized and employ directed cooling, where currents of cool water are circulated around the hotter portions of the engine, such as the exhaust valve seats (Figure 9.6).

A further development of pump circulation is the use of cross-flow radiators where the thermosyphon action is insignificant; a smaller number of longer tubes simplifies the construction of the wide but shallow radiators necessitated by a lower bonnet height.

The water-pump spindle is usually carried in grease-sealed ball races and has a spring-loaded carbon seal to prevent water leakage.

9.6   Pressurized radiator

Where a vehicle is operated under arduous conditions, particularly in mountainous regions, loss of water through boiling may occur. At 2000 m the atmospheric pressure falls to about 80 kPa and the boiling point to about 94°C.

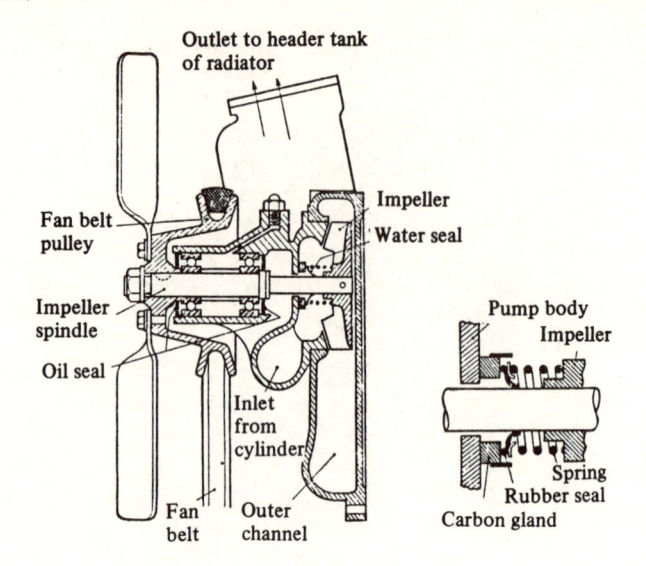

Figure 9.5 Centrifugal water pump

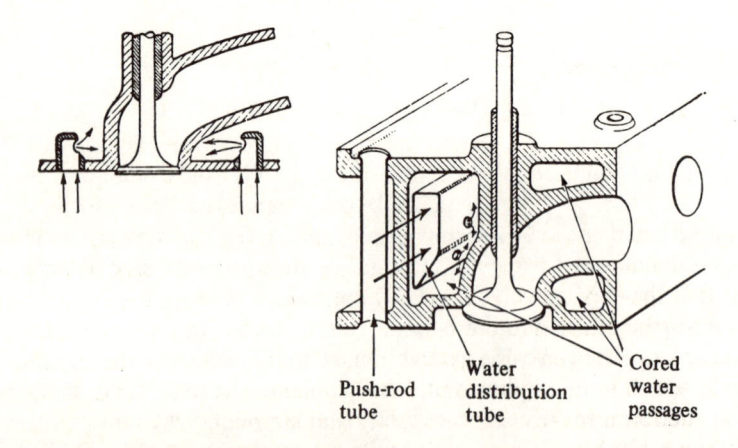

Figure 9.6 Directed water cooling

This loss of water can be minimized by using a pressurized cooling system, where the radiator is fitted with a spring-loaded outlet valve to maintain a pressure – created by the evaporation and expansion of the coolant as the temperature rises – usually between 30 and 130 kPa above atmospheric pressure. Each 10 kPa increase in pressure raises the boiling point of the coolant about 2°C. Since heat dissipation is propor-

The cooling system

tional to the temperature difference between the coolant and the outside air, the cooling efficiency can be raised and a smaller radiator fitted.

The engine operating temperature is raised to the optimum – around 100°C – and the pressurization maintains the water-pump efficiency at these temperatures. Loss of water through surge when braking is eliminated. An additional valve is necessary to prevent the formation of a vacuum during cooling. Provision is made to release pressure in the system through the vent before the cap can be removed, to prevent the possibility of scalding (Figure 9.7). The water will boil vigorously if the pressure is released at a temperature over 100°C.

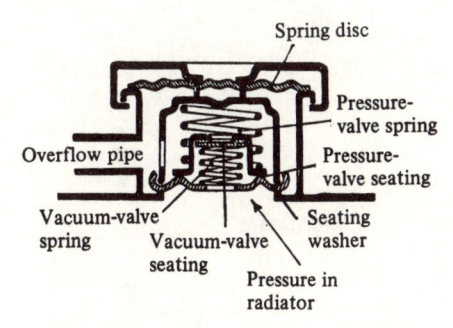

Figure 9.7   Pressurized radiator cap

9.7   Sealed system

The pressurized system is usually modified into the sealed system where any vapour or coolant vented from the radiator header tank is collected in an expansion tank placed in a cool position on the vehicle. This overflow is returned to the radiator by the partial vacuum created there on cooling. The vent pipe extends almost to the bottom of the expansion tank, which is maintained half-full of coolant. The sealed system allows a reduction in header-tank capacity and simplifies cross-flow radiator systems.

The pressure cap may be fitted either on the expansion tank or on the radiator – the expansion tank then being vented to atmosphere.

9.8   Thermostat

The aneroid thermostat consists of a metallic bellows partially filled with acetone, alcohol or similar volatile liquid controlling a disc valve in the water outlet from the cylinder block (Figure 9.8).

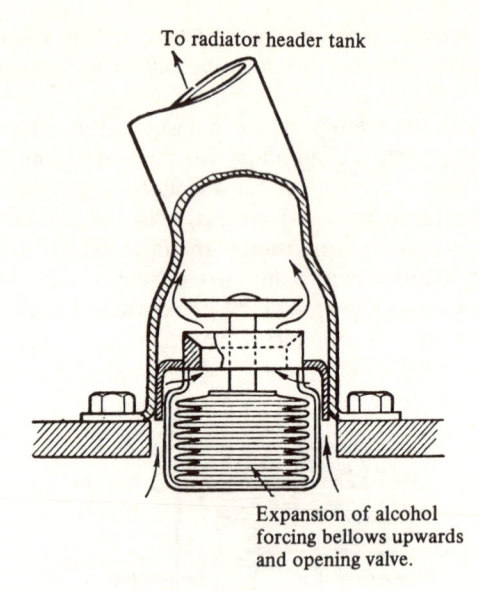

Figure 9.8 Aneroid-type thermostat

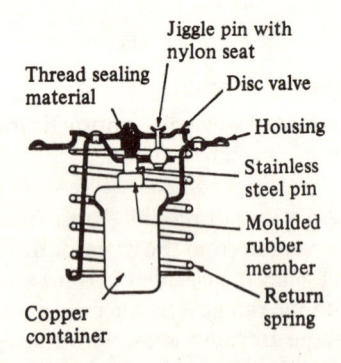

Figure 9.9 Wax-type thermostat (one half displaced 90°)

The wax-type thermostat, in which the expansion of melting paraffin wax deforms a moulded rubber membrane and displaces a stainless-steel pin from the rigid container, has the advantages of being insensitive to sudden temperature fluctuations or to the pressure in the system (Figure 9.9).

By controlling the flow of coolant through the radiator and regulating heat dissipation, the thermostat allows the engine to reach its optimum temperature as quickly as possible and then to maintain it. This results in reduced oil dilution, cylinder bore wear, petrol consumption and exhaust emission, and provides an adequate output from the interior heater.

9.9 Fan control

Rising road speed promotes increasing natural air flow through a conventional radiator, whilst unnecessary and increasing power is being lost driving the cooling fan.

A simple slipping clutch in the fan hub, using a viscous fluid between closely spaced members, or a more elaborate feathering-blade system, will limit the fan torque. The most common is an electrically driven fan, controlled by a thermal switch in the cooling system. This provides precise control and simplifies the radiator location.

9.10 Frost protection

Water has its maximum density – that is, the minimum volume for a given mass – at 4°C, and when cooled to 0°C it expands by about one-tenth. If the water in the cooling system is allowed to freeze, the expansion will often crack the cylinder block and other vulnerable parts; if the engine is turned, the impeller or water-pump drive will probably be broken. Occasionally the fortuitous expulsion of the casting core plugs may provide sufficient release of stress in the block casting.

If draining-off is adopted – usually impracticable when an interior heater is fitted – any water-pump or cylinder-block taps must be opened in addition to the radiator drain, and the pressure cap removed. With this method care is needed since, with the circulation restricted by the thermostat, the water in the radiator may be frozen by the cold current of air as the vehicle proceeds. No circulation can then occur and the water in the cylinder block may reach boiling point and steam be discharged, simultaneous freezing and boiling taking place owing to the poor conductivity of water. Blanking-off or the use of a radiator blind avoids this difficulty.

A heated garage, radiator lamps or a permanently fitted mains electrical immersion heater in the cooling system have the advantage of maintaining engine temperature, and are employed when vehicles may be needed for emergency services.

9.11 Antifreeze solution

A 25 per cent ethylene glycol solution will circulate down to $-12°C$ and a 50 per cent one down to $-37°C$. There is a further 13°C fall in each case before the soft ice hardens and expands. These solutions also raise the boiling point by 3°C and 9°C respectively.

Before using antifreeze solutions the cooling system should be drained and flushed and the joints checked for tightness. This is necessary since the solution has a cleansing and penetrating action and may either disturb accumulated sediment or provoke leakage.

The antifreeze proportions of a coolant can be determined using a hydrometer and attached thermometer. A chart provided with the instrument indicates the antifreeze percentage from the temperature-corrected density.

9.12 Maintenance

Air flow

The air passages through the radiator core, or on an air-cooled engine the cylinder fins, must be clean and the air flow must be unobstructed.

The fan belt, usually also driving the water pump and alternator, should allow some 20 mm deflection midway between the pulleys. Excessive tension causes wear of the belt and bearings. The belt must not 'bottom' in the pulley grooves or the wedging action will be lost.

Water

Hard water produces a limestone deposit in the water passages which, because of its poor heat conductivity, is likely to produce overheating (Figure 9.10). Soft water or rain water should be used, or alternatively potassium dichromate or some other corrosion inhibitor should be

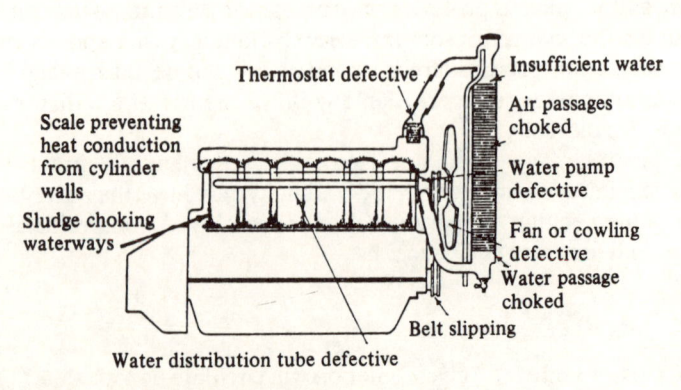

Figure 9.10 Causes of overheating

The cooling system

added to the water. Frequent renewal of the cooling water is not advisable as it may introduce more impurity.

Radiators should not be removed and stored during overhaul without flushing; accumulated sludge can solidify and permanently choke the core. Impaired circulation may be detected by the absence of a surge of water in the filler neck when the engine is speeded up, by an abnormal temperature difference across the radiator core, or by flow-testing the radiator.

When flushing is necessary, the reverse-flow method – removing the radiator hoses and thermostat and flushing up through the radiator and down through the block – is the most effective in dislodging sediment. For more vigorous action a solution of 1 kg of washing soda to every 10 litres of water can be used in the system, followed by draining and flushing.

Cold water should never be added to a hot engine since the sudden contraction may crack the cylinder head or block. Either hot water must be used, or the engine must be allowed to cool.

Thermostat

If the cooling system overheats or fails to attain its working temperature in the normal time, the thermostat must be suspect. When tested by heating in water, the thermostat should start to open within about 3°C of the marked temperature and be fully open within another 13°C.

Pressure cap

Pressure-cap action can be observed by the release of pressure when, observing safety precautions, it is eased on a hot engine. Testing requires a pump and gauge unit to ensure that the marked pressure can be attained and held for some ten seconds. The same equipment can be used with appropriate adaptors to pressurize and detect leakage in the cooling system.

Chapter 10
The fuel system

10.1 Petrol production

The principal fuel used in motor vehicle engines is petrol. This is a colourless, highly volatile liquid, obtained by distillation from crude oils from wells in various parts of the world. Petrols distilled from different crudes possess different characteristics depending on the proportions of paraffins, naphthenes and aromatics. All, irrespective of origin, are composed almost entirely of carbon and hydrogen, in the approximate proportions of 85 per cent carbon and 15 per cent hydrogen by mass.

This fuel is mixed with air, which contains 23 per cent oxygen and 77 per cent nitrogen by mass. When the spark occurs in the cylinder, combustion takes place and the oxygen combines with the hydrogen to form water (H_2O) and with the carbon to form carbon dioxide (CO_2).

10.2 Mixture proportions

For chemically correct, or stoichiometric, combustion ($\lambda = 1.0$, where λ (lambda) is the ratio of air supplied to air needed for complete combustion) the ratio of air to fuel should be approximately 15:1 by mass. This means that about 9 m^3 of air has to be mixed uniformly with 1 litre of petrol, and this is the function of the carburettor or fuel-injection equipment.

If maximum economy is required a weaker mixture (up to 20:1) will achieve this by ensuring that all the fuel is burned, but with a slight reduction in maximum power. If maximum power is sought a slightly richer mixture will ensure that all the oxygen is burned, but at the cost of much increased fuel consumption. A 6 per cent increase in power will involve an 18 per cent increase in petrol consumption. Mixtures weaker than 22:1 or richer than 9:1 will not burn in a normal engine.

The nitrogen of the air takes no part in the combustion process as outlined but is involved in exhaust emissions.

The fuel system

10.3 Octane number

One of the most important characteristics of a fuel is its tendency to detonate, and this is assessed by comparing it, in a variable compression-ratio engine, with a test mixture of iso-octane and normal heptane, two members of the paraffin series having respectively high and low antiknock values. If, at a certain compression ratio, the fuel has the same tendency to detonate as a mixture of, say, 85 per cent iso-octane and 15 per cent heptane, it has the octane number 85. The higher the octane number the greater is its resistance to detonation. Two variants in testing procedure result in the RON (research octane number) and MON (motor octane number); fuel-pump rating is usually the mean of the two.

10.4 Volatility

Volatility is a measure of the ease with which a fuel vaporizes. The constituents of petrol have boiling points ranging from 30°C to 200°C, and a high volatility from a proportion of the fuel is necessary for easy cold starting.

10.5 Calorific value

The calorific value – or *specific energy* – of a fuel is the total heat liberated when it is completely burned, and is expressed in J/kg, kJ/kg of MJ/kg. All petrol has much the same calorific value – about 46 MJ/kg. When mixed with air in the chemically correct proportion, fuels have a calorific value of some 4 MJ/m^3.

10.6 Blending agents

Benzole

Benzole is a hydrocarbon, obtained during the production of industrial gas from coal. It has a high octane number and is valuable as a blending agent – up to about 25 per cent of the total volume – to improve the qualities of the fuel.

Alcohol

Alcohols contain carbon, hydrogen and oxygen and are produced from vegetable matter, which (unlike minerals) is a renewable energy resource. Both the calorific value and the air/fuel ratio are lower than with the hydrocarbon fuels, resulting in a much increased fuel consumption, but the high octane rating of the alcohols makes them valuable blending agents; up to about 15 per cent by volume is used.

Alcohol has a much higher latent heat than petrol, and the quantity of heat required to vaporize it produces cool running and improved volumetric efficiency owing to the greater density of the colder charge. These advantages, together with the high octane rating, make methanol (when permitted) an excellent fuel for high-compression racing engines.

Tetra-ethyl-lead

The cheapest method of raising the octane rating of a fuel is by adding up to about 0.05 per cent by volume of tetra-ethyl-lead. A larger proportion can result in fouling of the plug insulators and binding of the exhaust valves in the guides. Environmentalists indict tetra-ethyl-lead as a toxic pollutant and press for its removal from fuel.

10.7 High-octane fuel

All the liquid fuels give much the same power output from a given cylinder capacity and compression ratio. The use of a high-octane fuel is advantageous in allowing the compression ratio to be raised without

Table 10.1 Petrol ratings

Petrol rating	Octane number (typical)	Compression ratio (typical)	Specific energy (MJ/kg)	Latent heat (kJ/kg)	Air/fuel ratio (mass)
**	90	7:1	46/47	330	15:1
***	94	8:1			
****	98	9:1			
Alcohol racing fuel	100/130	16:1	25	1050	8:1

causing detonation and so resulting in improved thermal efficiency, permitting either increased power, increased economy or an improvement in each of these. Table 10.1 gives details of some petrol ratings.

10.8 The mechanical pump

An eccentric on the rotating camshaft actuates the rocker of the pump, which depresses the diaphragm and so creates a depression in the

The fuel system

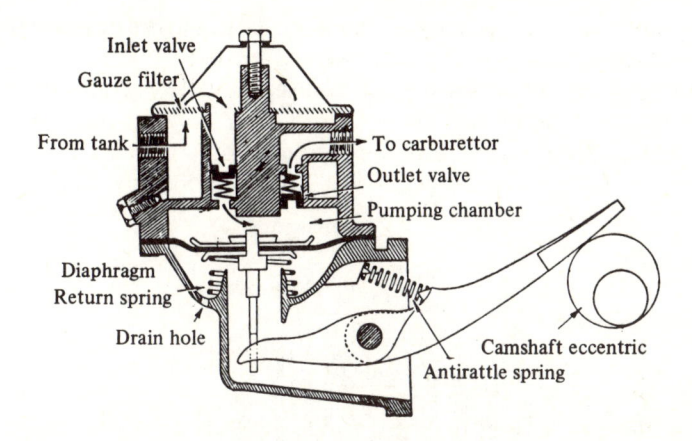

Figure 10.1 Mechanical fuel pump

pumping chamber (Figure 10.1). Under atmospheric pressure, petrol passes through the pipeline, filter and inlet valve into the pumping chamber. The return spring then raises the diaphragm, expelling the fuel through the outlet valve, pipeline and filter (if fitted) to the float chamber.

When the float chamber is full, float buoyancy closes the needle valve and the pressure in the pipeline and pumping chamber holds the diaphragm depressed against the tension of the return spring. The rocker lever is made in two parts; the outer portion moves idly on the eccentric under the pressure of the antirattle spring, and lost motion occurs between the two parts.

An external priming lever is frequently fitted to enable the diaphragm to be depressed by hand.

10.9 The electrical pump

The pumping-chamber action is similar in the electrical pump to the mechanical type. The pump body contains a solenoid which receives current via the ignition switch and a pair of contacts on the pump (Figure 10.2). The magnetic field created by the solenoid attracts an armature, or soft-iron disc, on the pump diaphragm, so causing the diaphragm to deflect and petrol to enter the pumping chamber.

At the end of the stroke a toggle mechanism operated from the diaphragm separates the pump contacts, the magnetic field collapses and the diaphragm is returned by a return spring, pumping fuel to the carburettor. The electrical circuit is then completed by the contacts and the action continues. When the float chamber is full, the pressure keeps the diaphragm displaced with the contacts open.

Electrical pumps are available to suit varying delivery rates, delivery pressure and mounting position.

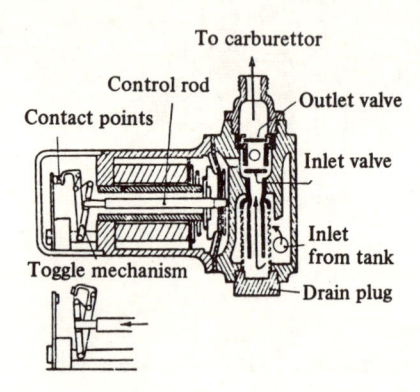

Figure 10.2 Electrical fuel pump

10.10 Maintenance

The fuel filters require cleaning every 10 000 km, and after prolonged service the diaphragm, valves and seats, and contacts require attention.

The correct delivery pressure of the pump – between 15 kPa and 35 kPa – is important as it determines the fuel level in the float chamber. The electrical pump can be bench tested for specified performance. On the mechanical pump, final pressure adjustment is made by gasket shims beneath the mounting flange, and a pressure test under operating conditions is necessary for accuracy.

A low mounting for the pump is desirable to reduce the suction lift. Neither the pump nor the pipeline must be situated near the exhaust pipe or manifold where the high temperature can evaporate the fuel and cause a vapour lock.

10.11 Air filters

The purpose of the air cleaner is to remove grit and dust from the hundreds of cubic metres of air consumed in each hour's running and so minimize cylinder-bore wear and oil contamination. The filter must be able to pass the required volume of air without restricting volumetric efficiency. The cleaning and replacement intervals should be modified according to the operating conditions.

Centrifugal separation

Vanes imparting a high-speed swirl – the cyclone action – can result in over 80 per cent of particles being centrifuged out of the intake air. This principle is applied as the first stage in many types of filter, particularly where arduous conditions prevail.

Paper filtration

Resin-impregnated paper, pleated into a multipointed star for the maximum surface area, provides both surface and depth filtration (Figure 10.3). Being compact these filters are valuable where space is limited and are widely used. Element renewal is required at 20 000–40 000 km intervals.

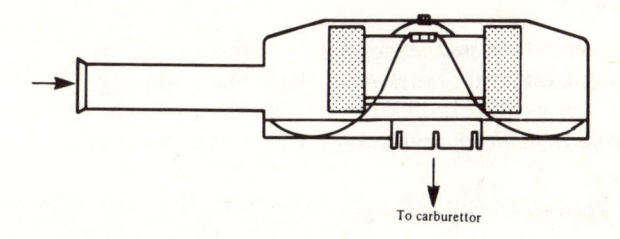

Figure 10.3 Pancake air cleaner

Oil-washed filtration

The intake air direction is reversed over an oil bath, projecting particles into the oil. The air then travels upwards through an oiled wire mesh which traps remaining material. The polluted oil gravitates down to the oil bath and is replaced by the oily mist carried on the air flow (Figure 10.4).

Intervals for cleaning and recharging with oil are determined by the usually severe conditions where this type of filter is employed.

Air silencing

Intake hiss is suppressed by the resonating-chamber effect of the filter air capacity and filter inlet tube tuned to attenuate the major frequency.

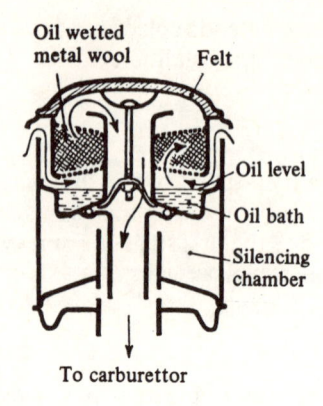

Figure 10.4 Oil-bath cleaner and silencer

Example 18

On test a petrol engine develops a brake power of 18.65 kW and consumes 2.84 litres of fuel during 20 minutes' running. If the fuel has a relative density of 0.78 and a calorific value of 46 MJ/kg, calculate the brake thermal efficiency and the specific consumption in litre per kWh.

$$\text{mass of fuel used per min} = 2.84 \times \frac{0.78}{20} \text{ kg}$$

$$\text{heat supplied per second} = \frac{2.84 \times 0.78 \times 46}{20 \times 60} \text{ MJ}$$

Now the power developed = 18.65 kW, and hence the work done per second = 18.650 J. We have

$$\text{thermal efficiency} = \frac{\text{work done}}{\text{heat supplied}}$$

$$= \frac{18\,650 \times 20 \times 60 \times 100}{2.84 \times 0.78 \times 46\,000\,000}$$

$$= 22\%$$

The volume of fuel used per hour = 2.84×3 litres, and the BP produced = 18.65 kW.

Hence

volume of fuel used per kW BP per hour $= \dfrac{2.84 \times 3}{18.65}$

$= 0.457$ litre

That is,

specific fuel consumption $= 0.457$ litre/kWh

Chapter 11
Carburation

The function of the carburettor is to control the quantity and proportion of the fuel and air entering the cylinders, to atomize the petrol into very small droplets and to vaporize this fine spray into a homogeneous combustible mixture.

11.1 The simple carburettor

The float chamber provides a constant level of petrol, just below the outlet from the jet, by means of the float and needle valve. An air hole allows atmospheric pressure to act on the petrol in the float chamber. A choke tube or venturi is fitted to increase the velocity of air over the jet (Figure 11.1). As the particles of air enter the venturi they must

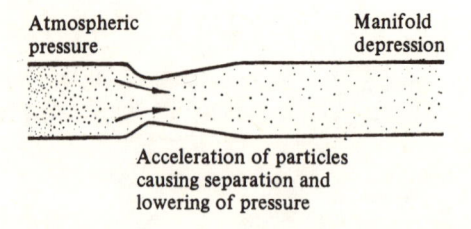

Figure 11.1 Choke-tube action

accelerate to pass through the smaller cross-section at a higher speed. During this acceleration the distance between each particle is increased, which corresponds to a fall in pressure; hence the venturi creates a greater depression over the jet.

The throttle valve regulates the quantity of mixture passing to the cylinders and therefore the IMEP which will be produced by the combustion of the charge.

The so-called simple carburettor (Figure 11.2) will provide a correct mixture for one air speed – or depression in the choke tube. Owing to

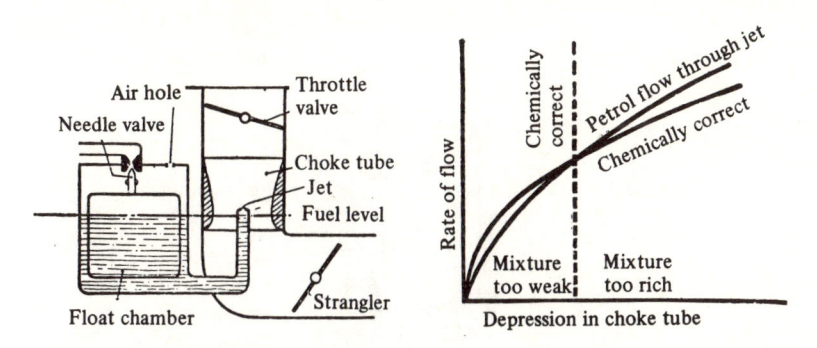

Figure 11.2   Simple carburettor

the different flow characteristics for petrol and air, this carburettor gives too rich a mixture at higher air speeds and too weak a mixture at lower speeds. To overcome this disadvantage and provide satisfactory carburation on a variable-speed engine, some form of compensation must be used.

11.2   Compensation

Air-bleed compensation

In this arrangement a well of petrol, fed from the main jet, is subject to the choke-tube depression. As the fuel level is lowered, air-bleed holes are progressively uncovered, thereby admitting extra air which reduces the depression acting on the jet and so prevents the increasing choke-tube depression from causing enrichment of the mixture (Figure 11.3).

Compensating jet

The main jet acts as in a simple carburettor and provides an increasingly richer mixture at higher air speeds. The submerged compensating jet feeds to a well open to atmospheric pressure. When this well has been emptied, which occurs at a comparatively small depression, the flow through the compensating jet depends only upon the constant head of petrol in the float chamber and not upon choke-tube depression. This constant flow of petrol, mixed with an increasing volume of air, results in an increasingly weaker mixture from the compensating jet at greater depressions (Figure 11.4).

By suitably balancing the main and compensating jets a compensated mixture strength can be obtained throughout the range of the instrument.

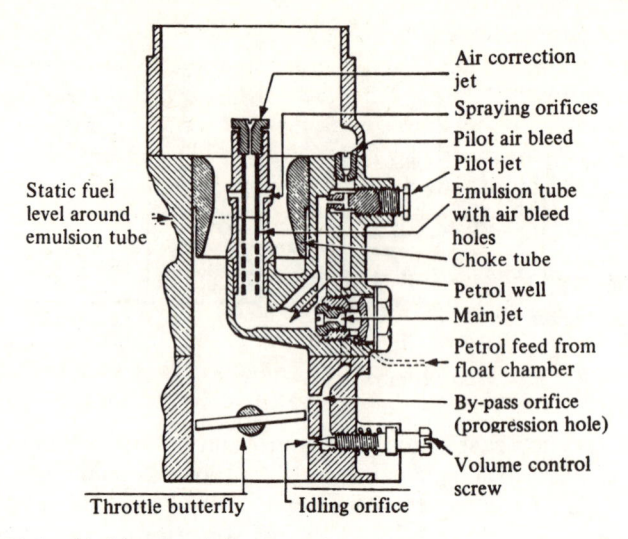

Figure 11.3 Air-bleed and slow-running arrangements

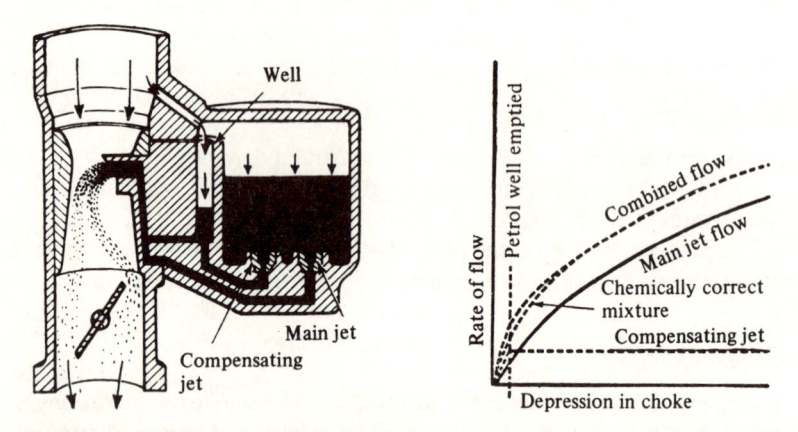

Figure 11.4 Compensating-jet action

In many instruments the compensating jet is used in conjunction with air-bleed compensation.

11.3 Dual characteristics

While 15 kg of air are required for the chemically correct combustion of 1 kg of average fuel, a weaker mixture (down to 20:1) gives the highest thermal efficiency by ensuring complete combustion of the fuel, and a richer mixture (up to about 13:1) gives the highest BMEP by utilizing all the air available. To meet both requirements the carburettor should therefore provide a weak 'economy' mixture on part-throttle opening and a rich 'power' mixture on full throttle, irrespective of the engine speed and therefore the air flow through the instrument. The carburettor has therefore to be designed to meet these dual characteristics.

By a suitable combination of the main and compensating jets, or the size and number of the air-bleed holes, *over-compensation* can be obtained where the mixture strength is richer at low and weaker at high air speeds, thus reversing the characteristics of the simple carburettor. This over-compensation provides the economy requirements for part-throttle conditions. In addition, the idling system is often fed from the petrol well, and when this is emptied the idling system becomes an air bleed.

For full-throttle condition, power enrichment can be obtained from a power jet mechanically opened by the final movement of the throttle lever. Alternatively, an economizer valve operated by induction-pipe depression against spring tension may be used (Figure 11.5). On full-throttle conditions the reduced depression allows the economizer valve to open an additional petrol jet or close an air bleed.

Dual-characteristic requirements and other provisions of carburation have to be modified by the overriding considerations of emission control legislation.

11.4 Starting devices

When starting from cold a smaller proportion of the fuel is evaporated and some of this will condense on to the cold surfaces during its passage into the cylinder. As a result a normal mixture of 15:1 would be too weak for combustion, and enrichment is required.

A strangler, spring loaded and mounted off-centre, or incorporating a spring-loaded valve, may be fitted (Figure 11.6). An interconnection opens the throttle to a fast idle when the strangler is operated. As the engine speed rises the increased depression overcomes the spring loading, opening the strangler or valve and admitting air to prevent over-choking.

Alternatively, a manually controlled starter carburettor can be fitted (Figures 11.7, 11.8). This is supplied with mixture from separate air holes and petrol jet, and feeds to a drilling on the engine side of the throttle valve. Reduction of the mixture strength as the engine warms up can be obtained by using a spring-loaded extra air valve, or may depend upon the emptying of a petrol well fed from the starter-carburettor petrol jet.

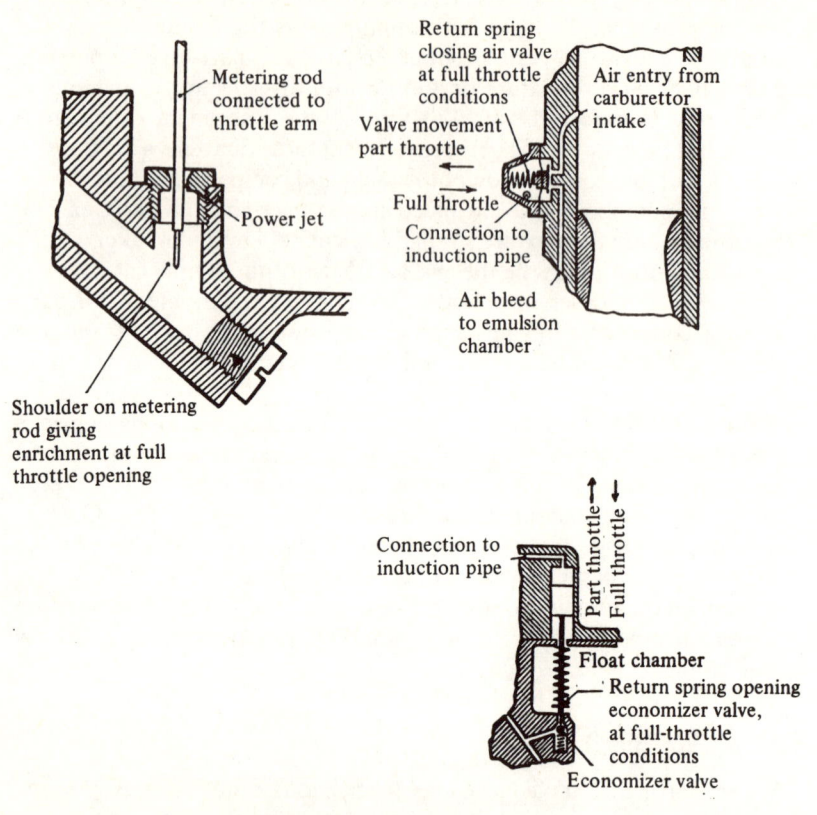

Figure 11.5 Full-throttle enrichment devices

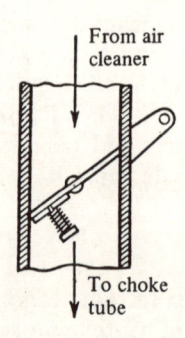

Figure 11.6 Strangler

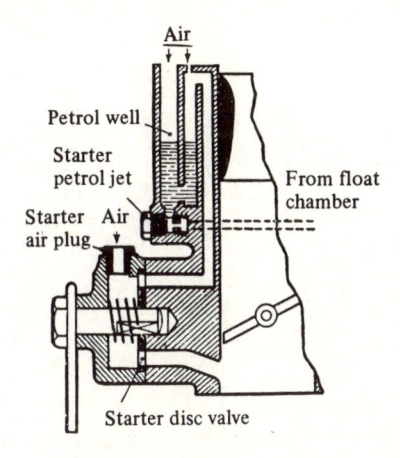

Figure 11.7 Starter carburettor: disc valve

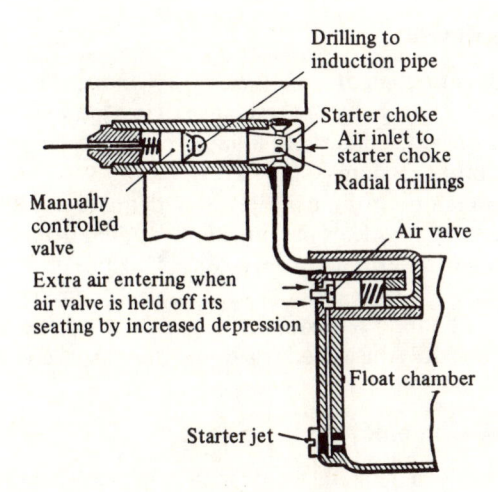

Figure 11.8 Starter carburettor: piston valve

Over-strangling results in oil dilution, abnormal cylinder-bore wear, reduced fuel economy and serious hydrocarbon emission. In addition, enrichment may prevent starting; in this case, ventilating the engine by turning with the throttle fully open and the starting device out of action will usually effect a cure.

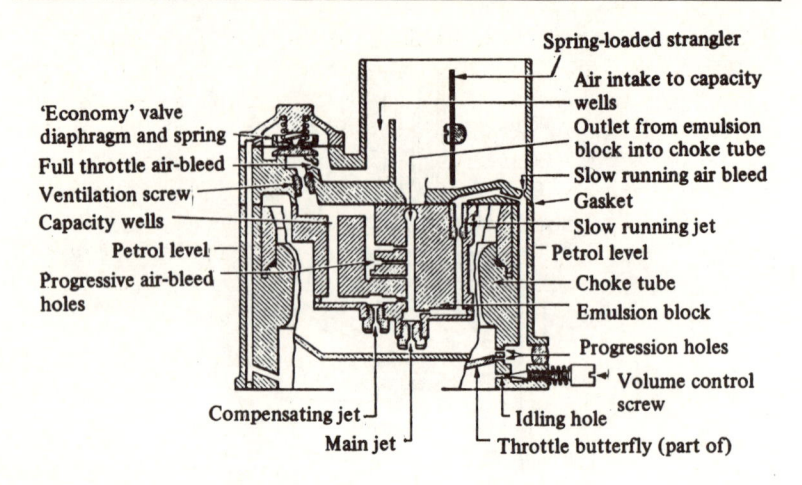

Figure 11.9 Slow-running compensation and economy arrangements (extended schematic view)

11.5 Idling devices

During idling, enrichment is necessary to compensate for the contamination of the charge by the residual exhaust gas and, where a large valve overlap is used, exhaust gas induced back into the cylinder by the high inlet manifold vacuum (Figure 11.9).

As the depression on the main jet is inadequate at idling speeds, a drilling on the engine side of the throttle valve, fed from a pilot jet and controlled by an adjustable screw, is used to provide the necessary mixture.

An additional idling adjustment is provided by the throttle stop screw, which controls the closed position of the throttle valve.

11.6 Progression holes

When the throttle is opened, the depression must be transferred from the idling outlet to the main jet. To assist in this transfer and prevent a weak mixture, progression holes, communicating with the idling system, are usually drilled near the edge of the partly open throttle valve.

11.7 Acceleration devices

Owing to the greater inertia of petrol compared with air, a sudden throttle opening is liable to produce a mixture too weak for combustion,

Carburation

resulting in a flat spot where the engine will not accelerate promptly.

With some carburettors, the well of petrol, since it is on the engine side of the main jet, is readily available to obviate the flat spot. Alternatively, an accelerator pump interconnected with the throttle valve may be fitted (Figure 11.10). The accelerator pump injects a fine spray of fuel into the choke area when the throttle is opened sharply; the spray can be prolonged by incorporating a spring in the operating linkage. The pump plunger has a limited clearance in its cylinder which allows the fuel to escape when the throttle opening is gradual and no enrichment is required.

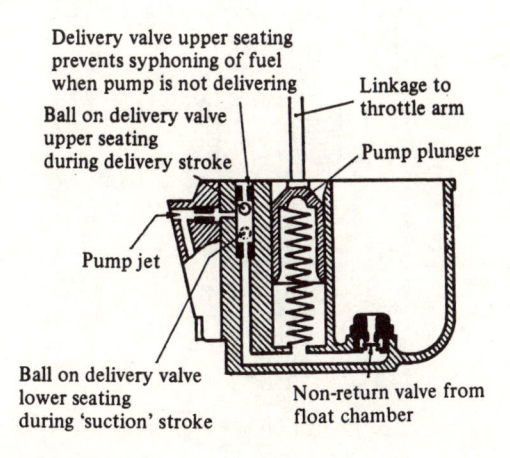

Figure 11.10   Accelerator pump: mechanical

In some cases a flexible diaphragm is used instead of a plunger for the accelerator pump. The membrane can be connected to the throttle linkage or operated by manifold depression acting on its outer face against spring loading; part-throttle, high-vacuum conditions fill the chamber, whilst the reduced vacuum on throttle opening allows the spring to eject the fuel.

11.8   Constant-vacuum carburettor

In this instrument the choke area is variable and is controlled by a movable piston subjected to throttle-chamber depression on its upper surface and atmospheric pressure below (Figure 11.11). The piston is lifted by the difference between these pressures and takes up a position of equilibrium where its weight and a light return spring is balanced by

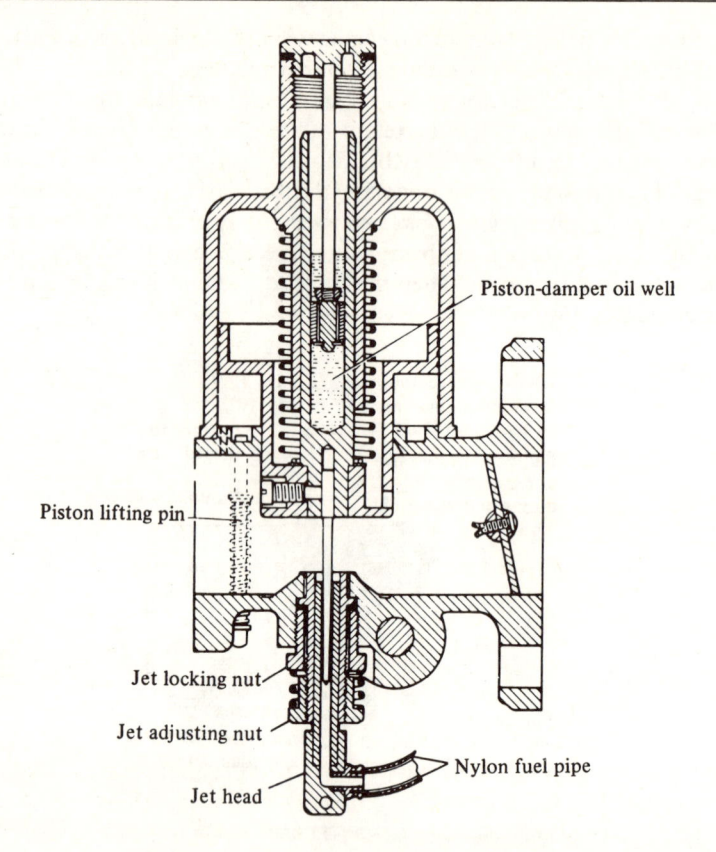

Figure 11.11 Constant-vacuum carburettor: piston seal

the constant depression in the throttle chamber – some 2.25 kPa. The piston therefore rises and falls as necessary to maintain this constant vacuum irrespective of engine speed or throttle position. If the throttle-chamber depression increases it will lift the piston, so enlarging the choke area and reducing the depression to its constant value, and vice versa.

This provides conditions similar to a simple carburettor having a constant depression and air speed over the jet, but the jet area must also be regulated to suit the varying choke area to give a constant mixture strength. This is accomplished by a correctly ground tapered needle attached to the piston and moving inside the jet. Differently ground needles can be used to vary the mixture characteristics.

The jet is movable and can be lowered to provide a richer starting

Carburation

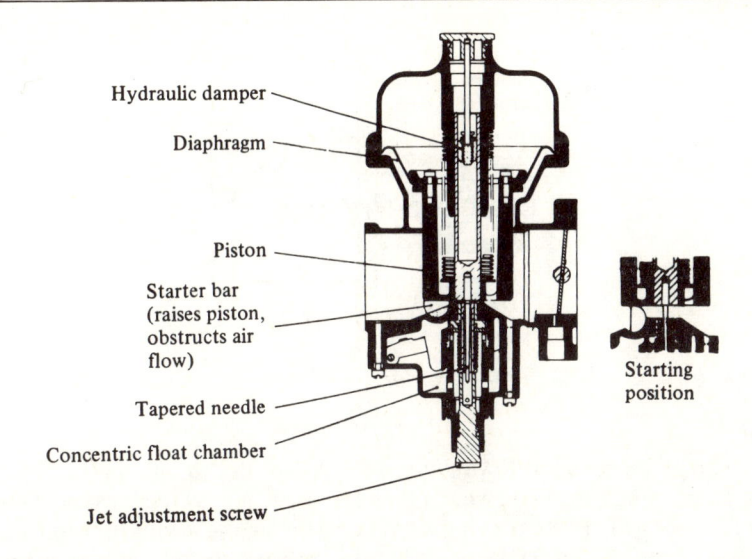

Figure 11.12   Constant-vacuum carburettor: diaphragm seal

Labels: Hydraulic damper, Diaphragm, Piston, Starter bar (raises piston, obstructs air flow), Tapered needle, Concentric float chamber, Jet adjustment screw, Starting position

mixture. An adjustment détermines the normal raised position of the jet and enables the mixture strength to be enriched or weakened throughout the range.

One-way hydraulic damping is employed to restrict the rise of the piston during acceleration and so produce a temporary enrichment of mixture strength, as in a simple carburettor. A light spring is fitted to supplement piston weight.

Correct centring of the jet assembly and free fall of the piston is essential, and is tested by using the piston-lifting pin when the engine is stationary. Idling-mixture strength can be checked by raising the piston 1 mm; only a very slight increase in engine speed should result.

Constant-vacuum carburettors, of piston- or diaphragm-sealed (Figure 11.12) types, are used with sidedraught or semi-downdraught manifolds; cornering, braking or acceleration inertia forces would affect a horizontal piston if used for a downdraught application.

11.9   Air velocity governor

In this simple governor the throttle valve turns freely on needle roller bearings on an offset throttle spindle connected to the accelerator in the usual manner. Below the governed speed the throttle valve is opened by the governor spring to an extent limited by a stop on the throttle spindle (Figure 11.13).

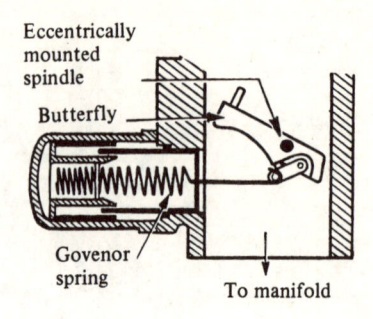

Figure 11.13 Carburettor governor

The pressure of air flowing through the carburettor acts on the larger portion of the throttle valve, tending to close it against the tension of the governor spring. By a suitable adjustment of the spring tension the engine can be restricted to a certain maximum air flow, since at this point the air pressure will overcome the governor spring and slightly close the throttle irrespective of the position of the accelerator.

11.10 Manifold systems

Sidedraught, semi-downdraught and downdraught inlet manifolds are used, the last two having gravity to assist, in varying degree, the mixture flow (Figure 11.14).

To assist vaporization the exhaust manifold usually contacts the inlet manifold to form a hot spot (Figure 11.15). A shutter, controlled by a metal thermostat, may be employed to deflect the exhaust gases around the hot spot when starting from cold. Afterwards the valve opens, preventing overheating of the charge and consequent loss of density and volumetric efficiency – always a potential source of reduced power when a hot spot is fitted.

The inlet manifold is sometimes separate, water-jacketed and heated from the cooling system; this is necessary with cross-flow engines (Figure 11.16). This system does not assist cold starting but ensures fuel vaporization at normal running temperatures. Figure 11.17 shows a typical exhaust manifold.

Mixture distribution and volumetric efficiency are dependent upon the inlet and exhaust tracts; interaction between the cylinders and between exhaust and induction during valve overlap is of considerable effect (Figure 11.18). Systems designed to utilize the negative exhaust-pressure wave at the commencement of induction and the ram effect of

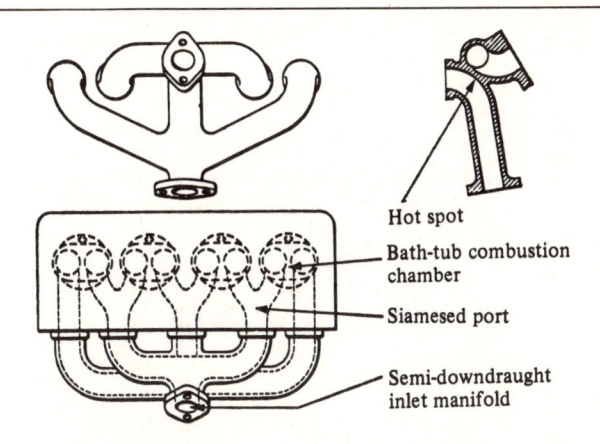

Figure 11.14 Manifold layout

- Hot spot
- Bath-tub combustion chamber
- Siamesed port
- Semi-downdraught inlet manifold

Induction manifold (downdraught)

Exhaust manifold

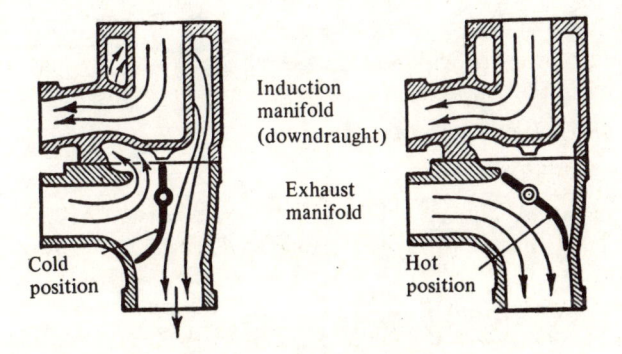

Cold position

Hot position

Figure 11.15 Thermostatically controlled hot spot

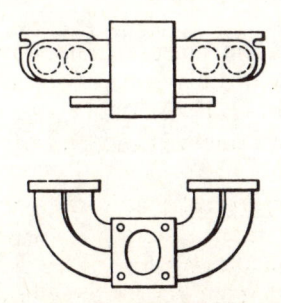

Figure 11.16 Type 1-4, 2-3 inlet manifold with water heating

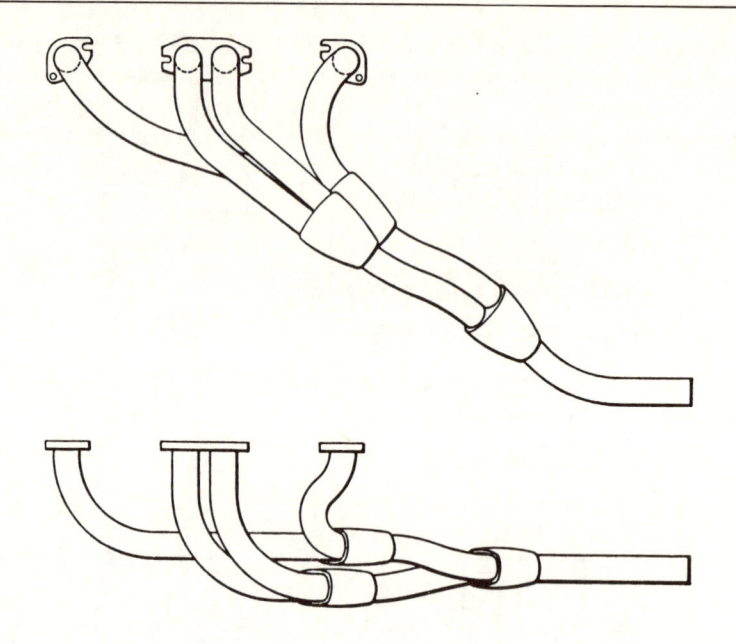

Figure 11.17   Type 1-4, 2-3 exhaust manifold

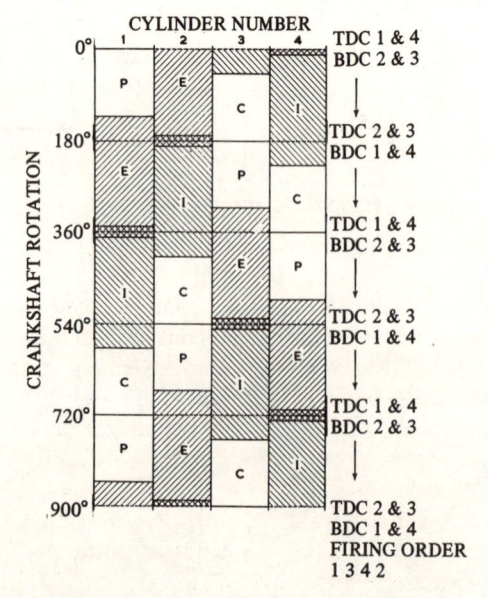

Figure 11.18   Induction-exhaust pattern for four-cylinder in-line unit

Carburation

inlet gas momentum at the end of induction can, for example, be very beneficial. On the other hand, detrimental effects can be produced; for example, siamesed ports may result in a loss of charge from a cylinder at the end of induction to one at the commencement; or the pressure wave from the opening of one exhaust valve may reach another cylinder where the exhaust period is almost completed.

11.11 Indications of mixture strength

A weak mixture at some part of the range is indicated by:

(a) Stalling – stopping from tick-over speed
(b) Flat spot when accelerating
(c) Lack of power
(d) Overheating
(e) Popping back in the carburettor; a weak mixture, being slow burning, may ignite the inlet gas as it enters the cylinder.

The effects of an over-rich mixture are:

(a) Hunting – a rhythmical surge in the engine speed when idling
(b) Excessive fuel consumption
(c) Black smoke, from unburned carbon particles, in the exhaust gas.

11.12 Emission control

Chemical pollution of the environment from vehicles is now controlled by legislation in many countries. This restricts the emission of carbon monoxide, CO; unburned hydrocarbons, generally labelled HC; oxides of nitrogen, generally labelled NO_x; and lead salts. The last are produced from the fuel additives tetra-ethyl-lead and tetra-methyl-lead. Emission control requirements have a substantial effect on engine design (Figure 11.19).

Reduction of CO and HC implies complete combustion and high thermal efficiency with the exhaust gas containing a maximum of carbon dioxide (CO_2) and water (H_2O). Unfortunately in a conventional engine a weak mixture can result in extended burning and high temperatures which, above 1650°C, foster the formation of NO_x. Special designs able to burn unusually weak mixtures can, however, lower the peak temperature by the excess air.

The combustion chamber should have a low surface/volume ratio and minimum clearances – e.g. on the top land of the piston – where the flame can be quenched, with the production of HC. Whilst the suppression of detonation favours the flame front moving towards cooler – and

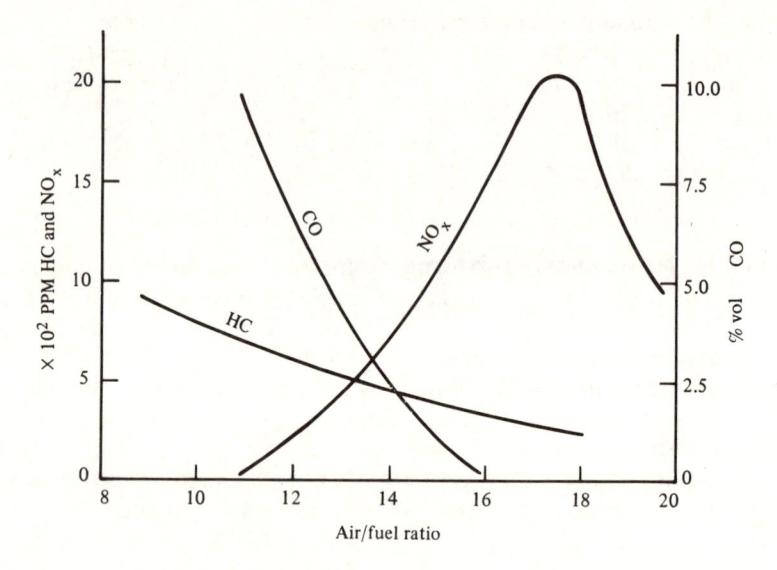

Figure 11.19 Exhaust emission

higher surface/volume ratio – quench regions, these can be an important source of HC.

Vigorous and controlled turbulence is needed to burn the weakest practicable mixtures. The intake air temperature should be controlled for optimum combustion. Modifications to the compression ratio, valve and ignition timing may be required. Fuel injection becomes increasingly attractive with the necessity for complex carburettor systems.

Where more stringent regulations apply, the use of injected air, and thermal or catalytic converters in the exhaust system, may be necessary.

11.13 Intake air temperature

Warm induction air provides latent heat for the vaporization of the carburetted fuel droplets. Excessive heating, however, lowers the density of the charge and the IMEP in the cylinders. The intake air temperature needs to be controlled to around 40–45°C.

At the air cleaner, hot air supplied from a shroud around the exhaust manifold can be apportioned to the ambient air by a flap operated by a bimetal strip, sensing air-cleaner or under-bonnet temperature. If required the valving can be servo-operated by an induction depression diaphragm controlled by the thermostat.

11.14 Automatic starting devices

An automatic starting device is required to preclude the possibility of gross emissions from the excessive use of a manual control. It must function at the coldest temperatures encountered and regulate the mixture to the minimum necessary for regular idling and satisfactory performance when driving away.

Automatic starting devices employ some of the following:

(a) A bimetal coil subject to engine coolant temperature. When cold this closes an air strangler, or retracts a tapered needle from a starter carburettor jet. As the temperature rises it renders these inoperative.
(b) A diaphragm or piston subject to manifold depression. This acts to open the strangler or insert the tapered needle into the jet so as to weaken the mixture once the engine starts. A temporary increase of manifold pressure during acceleration will enrich the mixture and prevent a flat spot.
(c) A stepped cam controlled by a bimetal coil sensing either coolant or carburettor ambient temperature. This regulates the weakening of the mixture by the induction depression device and also acts as a fast-idle throttle stop, allowing the throttle to close to normal idling speed as temperature increases.
(d) Where a starter carburettor feeds directly into the manifold the admission valve can be operated by a solenoid energized through the ignition switch. This circuit also includes a thermo-time switch where the bimetal element is subject to both coolant temperature and a heater coil. Should the engine fail to start, the heater coil will deflect the thermal strip after a predetermined time and prevent fuel flooding. Thermo-time devices are widely used to prevent excess fuel delivery if an engine fails to start.

11.15 Hot idling

The volume of the idling mixture may be only 2 per cent of the maximum carburettor flow, but irregular or incomplete idling combustion can cause substantial noxious emission.

Idling systems by-passing the throttle valve with separate adjustment for air and fuel can give a stricter control than those relying on the throttle-stop screw adjustment. The air inlets and outlets of these idling circuits are positioned to have equal pressures and prevent fuel delivery once the throttle opens.

The weak mixtures required for emission control can result in high-temperature combustion with the possibility of incandescence, producing running-on – auto-ignition – after the ignition is switched off. Some carburettors incorporate a solenoid-controlled valve to ensure that the slow-running fuel supply is cut off with the ignition to prevent this.

Exhaust-gas analysis is essential for tuning carburettors – especially the idling mixture – to meet emission legislation. After adjustment the controls are sealed, perhaps apart from a screw with minor effects on the idling volume.

11.16   Over-run conditions

Maximum depression occurs in the inlet manifold when the throttle is closed at speed – on the over-run – when the vehicle is driving the engine.

This high vacuum evaporates and draws off condensed fuel from the walls of the manifold, enriching the mixture and then leaving it over-weak and liable to cause spitting and popping back in the inlet and explosions in the exhaust. Both situations promote HC emission.

To relieve this high depression and provide sufficient mixture for regular firing, the throttle valve may have a small spring-loaded poppet valve fitted. Alternatively a throttle by-pass channel can be used with a spring-loaded diaphragm valve subject to manifold depression.

11.17   Air venting

On early carburettors the float chamber vented directly to the atmosphere. Later models admit clean air from the inner side of the air filter. This prevents differential pressure causing enrichment should the filter become clogged and restricts the emission of evaporated petrol fumes. With a hot engine these fumes may affect the slow running, and when the engine stops they may accumulate and prevent easy restarting. Some carburettors therefore employ a dual venting system where a throttle-linked valve transfers the float chamber venting to the atmosphere during idling.

Some legislation requires the restriction of fuel-tank HC fumes – usually by fitting a canister of carbon granules to the air vent.

11.18   Crankcase emission

The usual crankcase ventilation system admits clean air from the air filter into the overhead valve cover to circulate around the crankcase and be extracted, with hydrocarbon fumes and piston blow-by gas, into the induction system for burning in the cylinders.

With a variable-choke carburettor the crankcase can connect directly to the constant-vacuum chamber where the depression remains at around 2.25 kPa. With a fixed-choke instrument a crankcase ventilation control valve is necessary. With no depression, or with a pressure in the inlet manifold due to a backfire, the valve is closed. At maximum depression, during idling or over-run, the flow is severely restricted, and between these extremes the valve moves to allow adequate ventilation.

11.19 Modified jet systems

Any radial displacement of a centrally disposed needle in a jet alters the shape of the orifice and slightly affects the flow. Constant-vacuum carburettors now give the needle a bias from spring pressure in its holder, to maintain a light sliding contact with one side of the jet throughout its movement. The orifice, though varying in area, remains of constant shape and discharge coefficient.

A bimetal strip, immersed in the float-chamber fuel, can raise the jet as the temperature increases and prevent the lowered fuel viscosity producing a rich mixture. Alternatively an air bleed, controlled by a bimetal strip mounted on the carburettor body, can be used to by-pass the piston air valve and regulate the mixture strength with temperature increase.

Mixture variation with vehicle angle or movement is minimized by mounting the float chamber and jet concentrically.

An example of electronic control applied to the constant-vacuum carburettor is the use of a microprocessor, receiving signals of engine and ambient temperature, engine speed and accelerator position. These are computed and used to operate a stepper motor, moving a cold-start enrichment valve and a fast-idle throttle stop to precise positions and also to energize a solenoid fuel cut-off valve.

Starting, cold and hot idling and fuel supply on the over-run are controlled, with considerable advantages in emission control and fuel economy.

11.20 Fuel injection

Advantages

The inherent advantages of metering and distributing fuel to each cylinder by injection become more significant as emission control and economy demand carburettors of increasing complexity.

Inlet-port injection eliminates the carburettor choke restriction and the need for air pre-heating and allows an inlet-manifold design concerned solely with volumetric efficiency. Thermal efficiency is improved by the accurate metering of fuel to load and speed conditions, the elimination of the enrichment used to blanket uneven cylinder distribution, and the avoidance of fuel deposition in a cold inlet tract.

As in other applications, the electronic facility for monitoring and computing a large range of signals offers sophisticated control that is more effective and cheaper than mechanical alternatives.

Layout

In a typical layout, an electric pump feeds a fuel rail supplying solenoid-actuated needle-valve injectors directed at each inlet-valve throat. As a safety precaution, the pump electrical supply, after initial priming, continues only whilst the engine is running.

The rail injection pressure – some 250–300 kPa – is controlled by a spring regulator with a connection from its diaphragm to the inlet manifold. The differential between fuel and manifold pressures therefore remains constant so that the injected quantity is determined by the duration of injection.

The digital memory of the electronic unit is programmed for basic injector pulse periods for a range of engine speeds and loads. When the ignition is switched off it instigates a prescribed power-down programme and also retains certain data of the engine operating conditions.

An air-flow sensor of the hot-wire type compares the resistance of a sensing wire cooled by the intake air – reacting to temperature, velocity and density – with a compensating wire subject only to the air temperature, and provides a signal of mass air flow. This is modified by the main inputs of rail-fuel temperature, engine-coolant temperature, engine speed, throttle position and movement and road speed. Subsidiary modifying signals can also be applied.

Operation

All the injector solenoids are triggered, from an ignition signal, to open the needle valves simultaneously and inject half the required quantity of fuel at each engine revolution.

When starting from cold – at cranking speed – the number of injections per cycle is doubled to allow the extra fuel to be supplied. During the warm-up period the fuel requirements are accurately metered. A throttle by-pass, with a stepper-motor-controlled valve, regulates the mixture supply to provide a stable low idling speed irrespective of changes in engine resistance and load.

The throttle-valve movements are electrically sensed for both position and rate of change. Dual characteristics are provided, with an increased mixture strength for acceleration and full-load open-throttle 'power' conditions, and a decreased strength for declerating and light-load 'economy' requirements.

Fuel injection ceases on the over-run in accordance with predetermined criteria, and also when a prescribed maximum rev/min is reached. Hot auto-ignition run-on is inherently eliminated.

Typically with fuel injection the carbon monoxide in the exhaust gas will be limited to about 1 per cent – about half that obtainable with a sophisticated carburettor. A further development is the use of a sensor to monitor the oxygen content of the exhaust gas and thus assess the air/fuel ratio. By a feedback circuit this information can be used to modify the injection pulses.

Chapter 12
The compression-ignition engine

12.1 Operation

The four-stroke compression-ignition (CI) engine has a compression ratio of 16:1 to 24:1 and raises an induced charge of air to a pressure of 3–5 MPa and a temperature of 500–800°C at the end of the compression stroke. The fuel used has a self-igniting temperature of some 400°C and therefore ignites when injected into the combustion chamber at an injection pressure of 10–25 MPa.

12.2 Comparison of compression ignition and spark ignition

In the petrol engine a homogeneous combustible mixture, supplied by the carburettor, is ignited by the sparking plug. In contrast, on the compression-ignition engine the thorough mixing together of the fuel and air for complete combustion must be accomplished within the cylinder during some 35–40° of crankshaft rotation. This presents difficulty, and is accomplished by using a penetrating spray of atomized (finely divided) fuel and considerable turbulence of the compressed air in the cylinder. Even so, only some 75–80 per cent of the air supplied can be utilized, restricting BMEP and power, whereas the spark-ignition engine can consume virtually all the air induced.

Compared with the petrol engine the high combustion pressure necessitates a sturdy, heavy and costly construction. Inertia loadings from the more massive reciprocating parts limit maximum rev/min, and even with extensive light-alloy usage the power/weight ratio is lower. High-pressure fuel injection has always required specialized and costly equipment. Starting may present more difficulty owing to the high compression pressure and the heat required for fuel ignition. Idling is less smooth, acceleration is slower, maximum rev/min are lower and imperfect combustion will result in lower BMEP and obnoxious exhaust fumes. Fuel cleanliness is essential, and fuel handling less acceptable, whilst running-out may involve venting the injection system. Engine oil of a higher specification is needed.

The compression-ignition engine

On the other hand, the higher compression ratio gives increased thermal efficiency – over 30 per cent, compared with some 25 per cent for petrol units – and thus increased economy. Carbon, beyond a superficial coating, does not form when the engine is operating correctly. Exhaust temperatures are lower owing to the high expansion ratio. The fuel has a much reduced fire risk, is cheaper and has a considerable tax advantage in some countries. Engine torque at low speed is generally greater.

The characteristics of later-generation spark-engine-derived light high-speed CI units differ considerably from the earlier image of commercial diesel engines. Typically these units of 1.5–2.5 litre produce some 25 kW/l at around 4500 rev/min and maximum torque at under 3000 rev/min. Turbocharging will increase the output to 30 kW/l without loss of economy. A comparative petrol unit will produce 40 kW/1 at 6500 rev/min with maximum torque at 4000 rev/min.

12.3 Two-stroke CI unit

In a typical system air is supplied under pressure from a supercharger or turbocharger to ports in the cylinder uncovered as the piston

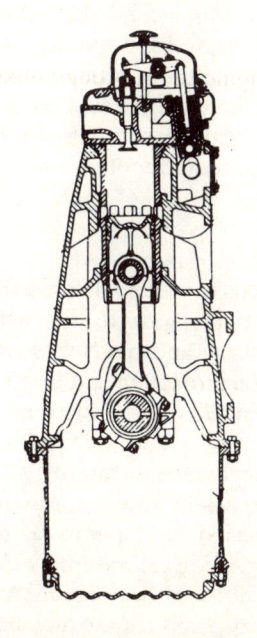

Figure 12.1 Two-stroke uniflow for pressure charging

approaches BDC, and uniflow scavenging takes place through exhaust valves in the cylinder head. Near TDC the compressed air receives injected fuel to provide the power stroke. Excess air, up to some 40 per cent, aids heat dissipation, whilst the charging pressure acts to reduce the delay period (Figure 12.1).

Each upward stroke is compression and, if the inertia forces at the operating speed do not exceed the gas load, the connecting-rod stress and bearing forces are not reversed as in the four-stroke unit. The torque is smoother but the complication is equivalent to the four-stroke unit and fuel economy is generally less favourable. Power outputs of 35 kW/l and BMEP over 1000 kN/m^2 have been obtained. For road transport the initial advantage of the pressure-charged two-stroke CI unit – greatly increased power for a given size and weight – has been reduced by the wide adoption of pressure charging on four-stroke units.

12.4 Three phases of combustion

Three phases of CI combustion are recognized. In the first or delay period, injection commences but there is an ignition lag where the fuel accumulates before burning. This may occupy about 10° of crankshaft rotation, but will be extended during idling owing to the lower temperatures and reduced fuel. In the second phase the flame develops and the combustion of the injecting and accumulated fuel produces an undesirably rapid rise in pressure with the characteristic diesel knock. During the third phase the remainder of the fuel is burned as it sprays into the chamber. Reduction of the delay period is sought by fuel quality, engine design and operation to obtain smoother running.

12.5 Fuel

When petrol (relative density around 0.74) has been distilled from crude oil, kerosene for gas turbines (some 0.78) is next obtained, followed by gas-oil (about 0.84). This is the main constituent of road vehicle diesel fuels, e.g. derv, diesolene. Heavier grades of fuel are used for large marine and stationary installations, where heating before injection is required.

Comparable with the octane rating of petrol, diesel fuel has an engine-test-based *cetane* rating which represents its readiness to ignite, so reducing the delay period and minimizing diesel knock. Apart from refining, amyl nitrite and ethyl nitrate may be added to improve the cetane number of the fuel. *Diesel index* is a simpler, less comprehensive assessment of the ignition quality of the fuel, based on fuel analysis.

12.6 Direct injection

With direct or open-chamber injection the compression ratio is relatively low (some 16:1) (Figure 12.2). An angled and sometimes masked inlet valve is used to impart swirl turbulence to the incoming air, which persists throughout the compression stroke (Figure 12.3). Near TDC a contracting squish area, between the rising piston and the cylinder head, violently projects the trapped air into (in current practice) a recess in the piston crown (Figure 12.4). A high injection pressure of some 25–30 MPa is employed to produce a hard penetrating spray from several minute injector nozzle holes into this turbulent air mass.

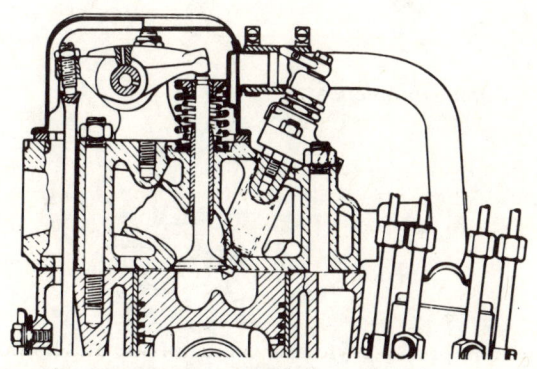

Figure 12.2 Direct injection

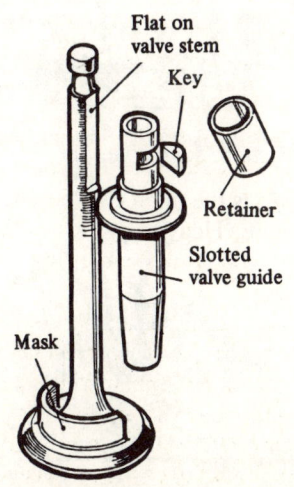

Figure 12.3 Swirl turbulence: masked inlet valve

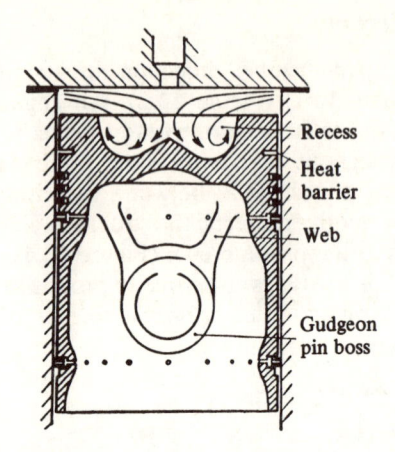

Figure 12.4   Squish turbulence

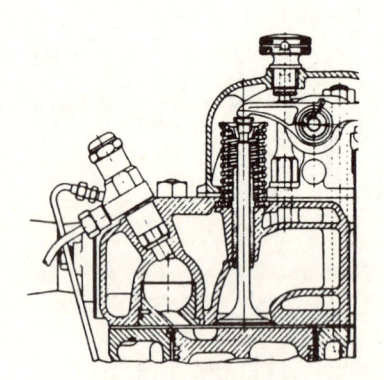

Figure 12.5   Indirect injection: swirl chamber

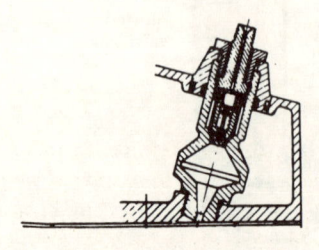

Figure 12.6   Indirect injection: pre-chamber

12.7 Indirect injection

A softer spray from a lower injection pressure, of some 10–15 MPa, takes place in a pre-chamber or (in current practice) a swirl chamber, into which the air is forced at high velocity by the rising piston (Figures 12.5, 12.6). This turbulence within the chamber promotes good combustion, and the expanding burning mass then projects from the chamber to promote turbulence and combustion within the cylinder.

12.8 Comparison of direct and indirect injection

Compared with direct injection, the more complete combustion obtainable with indirect gives smoother running, a higher BMEP and a cleaner exhaust with varying qualities of fuel. Since the higher rotational air speeds developed during compression are maintained in proportion to engine rev/min, higher-speed units – including car engines – are usually of this type. The pintle-type nozzles that can be used with the swirl chamber are also less prone to blockage.

However, cooling – due to the greater surface/volume ratio and air flow into the cell – results in lower thermal efficiency and higher fuel consumption. Compression ratios up to 24:1 may be needed, and an auxiliary heating device for cold starting is usually necessary.

12.9 Fuel supply

A piston or diaphragm low-pressure lift pump, operated from an eccentric on the engine or injection-pump camshaft, receives fuel from the tank through a primary filter and delivers via secondary filtration to the injection pump (Figure 12.7).

The working clearances between the mated parts of the high-pressure injection equipment – in-line pump plungers and barrels, distributor-pump rotor and hydraulic head and injector valve and body – are extremely small (about $1\,\mu m = 0.001$ mm). The efficiency and life of the equipment therefore depend almost entirely upon cleanliness of the fuel.

12.10 Filtration

Primary arrangements – on the intake side of the lift pump – usually comprise a sedimenter with the inflow and outflow arranged to allow the denser water and solid particles to settle in the base bowl. A gauze or nylon mesh strainer will remove material larger than about 50 μm but may cause wax blockage in low temperatures.

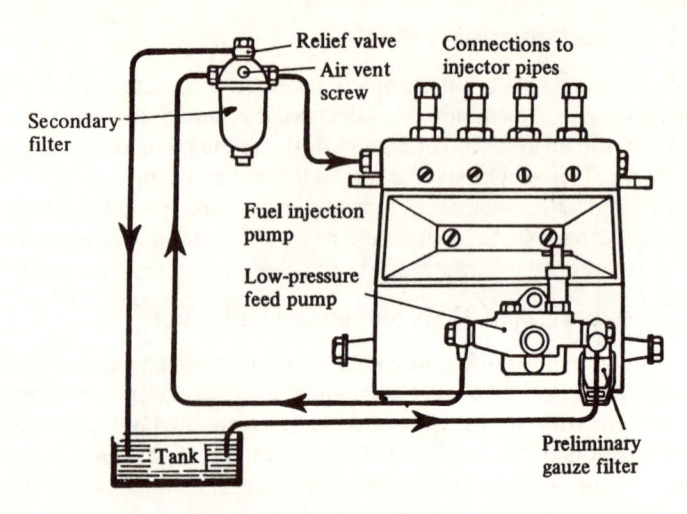

Figure 12.7 In-line injection-pump fuel layout

Secondary filtration has available the outlet pressure of the lift pump – some 35–100 kPa – and resin-impregnated paper is the preferred filter medium, removing particles larger than about 5 μm. This material may be radially pleated for inward cross-flow filtration – as in the usual lubricating-oil filter – or may be in a spiral V-shaped coil for up or down flow filtration. The latter is used in agglomerators where the water particles agglomerate into large droplets and settle to the base bowl. In some applications twin filters are used in series, the second operating at lighter duty and requiring less frequent changing.

As diesel fuel contains elements that commence forming wax particles below about −5°C, a fuel heater (battery powered) may be interposed in the secondary filtration.

12.11 Injection pumps

The injection pump must provide the exact quantity of fuel, precisely timed, that is required in each cylinder for the particular engine conditions. The injection requirements on small high-speed engines are particularly exacting, since the injection period will decrease to perhaps 0.001 s at maximum speed, and at idle the injected quantity will be reduced to a few cubic millimetres of fuel. For these engines the distributor-type pump is usual, whilst larger engines will employ the in-line unit.

Four-stroke injector pumps are driven by timing gears, chain or toothed belt at half the crankshaft speed.

12.12 In-line injection pump

The pump contains a separate plunger, barrel and delivery valve, with seating, for each cylinder. The plungers have constant stroke, being raised by cams on the pump camshaft through roller tappets and returned by springs (Figure 12.8).

A longitudinal fuel gallery feeds two radial drillings in each barrel – the inlet and spill ports, which may be neither level nor opposite. The fuel fills the space above the plunger and also a straight-cut groove in the side of the plunger – the so-called helix – via a central and radial drilling (Figure 12.9).

As the plunger approaches its maximum upward velocity it cuts off the uppermost port and the fuel is pressurized and forced through the delivery valve into the injector pipeline. This continues until the upper edge of the helix reaches the spill port, allowing the remaining fuel to escape. This spill point is determined by the angular position of the plunger in the barrel.

An arm on the lower end of each plunger slides in a vertically slotted control fork mounted on a longitudinal control rod (Figure 12.10). Movement of the control rod swings all the plunger arms and rotates

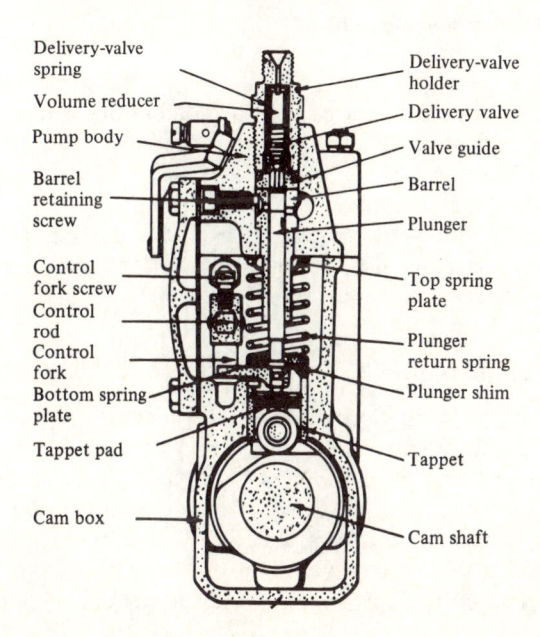

Figure 12.8 In-line injection pump

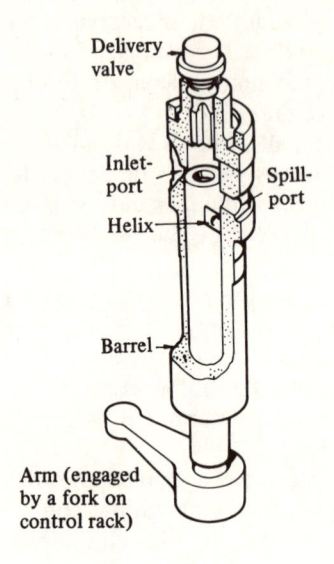

Figure 12.9   Pumping element

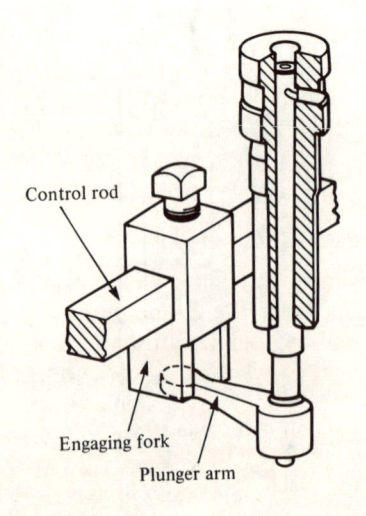

Figure 12.10   Fuel control

The compression-ignition engine

the plungers, so controlling the volume of fuel delivered to each injector from zero to the maximum, when the control rod contacts the maximum-fuel stop. The accelerator is connected, through the governor, to the control rod.

Small in-line injection pumps require periodical change of the lubricating oil as fuel dilution occurs. Larger pumps share the engine lubrication system.

Delivery valve

The spring-loaded non-return delivery valve is guided by a fluted shank, and carries below the conical seating face a closely fitting unloading collar. Before delivery can commence, this piston portion must be

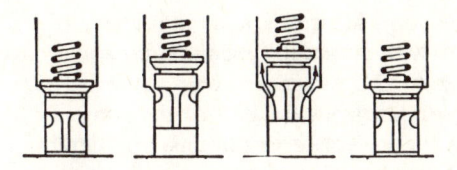

Figure 12.11 Delivery-valve action

raised clear of the seating. When delivery ceases, closure of the valve increases the pipeline capacity by the displacement of the piston (Figure 12.11).

This pressure reduction inhibits pressure waves in the pipeline adversely affecting the injection process, and allows the nozzle valve to snap shut and cut off the spray without dribble.

Phasing, calibration and maximum delivery

In order that each cylinder shall have an identical injection timing it is essential that the elements of the in-line pump commence to inject at exactly correct intervals of pump camshaft rotation – every 90° for four-cylinder or 60° for six-cylinder pumps running at half crankshaft rev/min. The precise setting of these phase angles, or phasing, is obtained by shim adjustment between each tappet and plunger, so regulating the cut-off point.

The pump must also be calibrated to ensure that each element delivers the same volume of fuel at any given position of the control rod. This is accomplished by an adjustment of the control fork on the control rod – a spill-point adjustment.

Maximum fuel delivery is adjusted by a stop limiting the travel of the control rod. These adjustments require appropriate servicing equipment and are carried out after pump overhaul.

12.13 Distributor-type injection pump

In this pump a single pumping element feeds to the various cylinders in turn, to a maximum of six (Figure 12.12).

A central steel rotor carries two opposed pump plungers which are operated by the internal lobes on a stationary cam ring. A central passage and radial ports in the rotor place the pumping chamber alternately in communication with a metering supply port and injector pipe delivery ports in the hydraulic head within which the rotor operates. These two members form a mated assembly, being manufactured to very fine limits.

In the usual layout, fuel from the tank is raised by a separate engine-driven lift pump with preliminary sedimenter and is passed through the main fuel filter to the injection pump, entering first a sliding-vane transfer pump driven from the end of the rotor. A regulating valve, situated within the end-plate, maintains a predetermined relationship between transfer pressure and pump rev/min by metering some of the output back to the inlet side of the transfer pump. When priming the pump the regulating valve also enables fuel to by-pass the stationary vanes.

From the transfer pump the fuel passes through a drilling in the hydraulic head and an annular groove around the rotor to the metering valve, which is controlled by the throttle lever and either a mechanical flyweight or a hydraulic governor operating at transfer pressure.

When one of the charging ports in the rotor registers with the metering port in the hydraulic head, the pump plungers are displaced outwards to an extent dependent on the volume of fuel that can enter; this is determined by the transfer pressure, the position of the metering valve and the time the ports are aligned.

Further rotation brings the cam rollers into contact with the cams; the plungers are driven inward, delivering the fuel through the distributing port in the rotor and one of the outlet ports in the hydraulic head to an injector pipe. The contour of the cam lobes relieves pressure at the end of the injection cycle and avoids dribble from the injectors.

The point at which the cam rollers contact the cam lobes varies slightly according to the plunger displacement, giving a slightly retarded injection timing under light load conditions. By providing cam-ring movement under hydraulic control this effect can be obviated, and in addition a retarded timing for starting and progressive advance with engine speed can be incorporated.

The compression-ignition engine

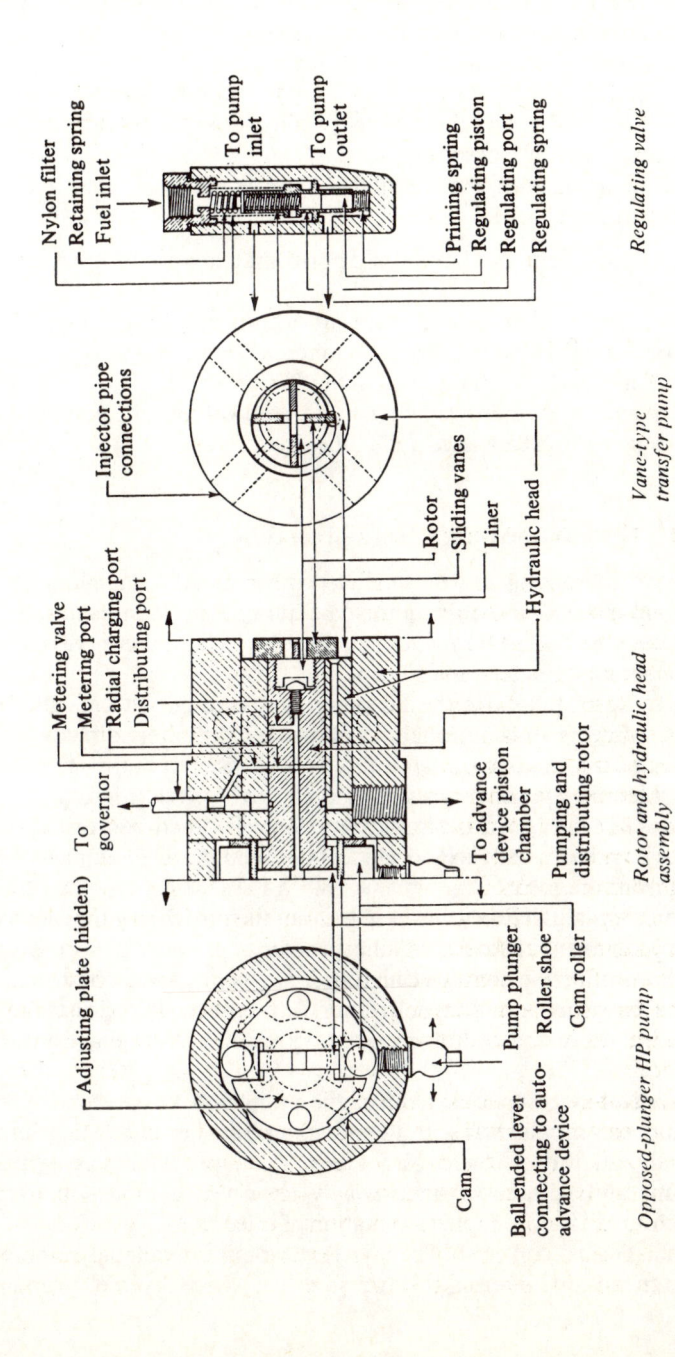

Figure 12.12 Distributor-type four-cylinder injection pump

The maximum outward displacement of the plungers and consequently the maximum fuel delivery from the pump is limited by projecting lugs on the roller shoes contacting eccentric slots in adjusting plates located on each side of the pumping section. Rotational adjustment of these plates by the makers or during overhaul permits a wide range of maximum fuel settings.

The advantages of the distributor-type pump compared with the in-line pump are:

(a) Compact, light, not limited to a horizontal mounting, and relatively cheap.
(b) Suitable for high rev/min; no reciprocating plungers and springs.
(c) Correct phasing and calibration inherent in the design.
(d) Self-lubricated by filtered oil under low pressure.
(e) Advance and retard can be incorporated without the complication of a centrifugal advance unit in the drive.

12.14 Distributor-pump developments

With the increasing use of smaller, higher-speed CI engines, many developments of the distributor-type pump are manufactured and licensed by Lucas CAV (as with petrol fuel injection, there are only two significant manufacturers in Europe).

On some installations the lift pump is dispensed with and the fuel passes from the tank, through a sedimenter and filter, directly to the transfer pump. A hand primer is fitted on the filter or in the line.

A solenoid-operated piston fuel cut-off to the metering valve provides key operation and instant stoppage. Toothed-belt-drive pumps are fitted with two ball races to support the belt load and eliminate radial thrust from the rotor.

Spring-loaded delivery valves in the outlets from the hydraulic head – in conjunction with the cam-lobe profiles – provide a more precise control of injection timing at high speed, give accurate control of the residual pressure in the injector lines, and ensure snap closure of the nozzle valves at the end of injection, precluding dribble and possible smoke.

A four-plunger development of the pump – with a second pair of plungers communicating with the pumping chamber and acting simultaneously on the adjacent cam lobes – can provide increased fuel pressure and volume or alternatively improved combustion from a higher rate and more rapid termination of injection.

Sophisticated control of the injection timing is available automatically from suitable internal fuel pressures or, where desired, by manual

The compression-ignition engine

control. This includes an automatic starting retard at cranking speed and advance under light load as well as the normal advance with speed. To ensure more stable running after a cold start, an advance with increased idle speed can be provided from an engine temperature thermocapsule or by manual control.

12.15 Governors

Compression-ignition engines generally use an all-speed governor, where the pump control rod, or the metering valve on distributor-type pumps, is operated only through the governor (Figure 12.13).

Depressing the accelerator increases the force of the governor spring, which tends to increase delivery, against the centrifugal force of the revolving governor mass, which tends to reduce delivery. At any pedal position the engine speed will increase or decrease until the centrifugal force is in equilibrium with the set tension of the governor spring. An increase in load will produce a fuel increase from the pump to stabilize the engine speed at the chosen rev/min (Figure 12.14).

The governor settings also provide a stable idling speed and limit maximum rev/min. A separate shut-off lever is usually fitted to in-line pumps to return the control rod from its idling position to that of 'no delivery'.

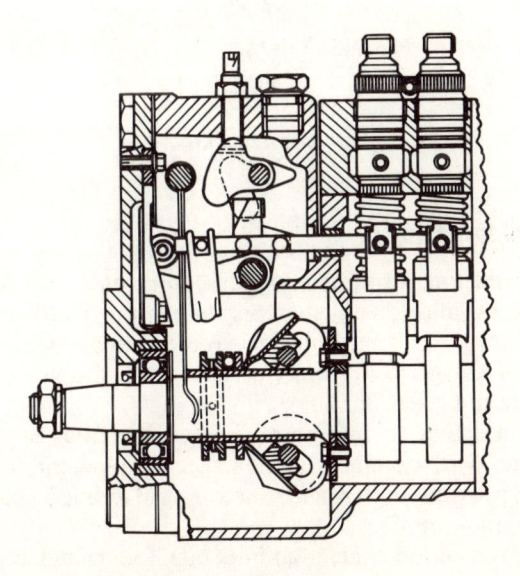

Figure 12.13   All-speed governor: leaf spring

The distributor pump can employ a simple hydraulic governor, where transfer-pump pressure, speed dependent, is used instead of the centrifugal force of revolving masses. On the distributor pumps used for light vehicles a two-speed mechanical governor may be used instead of the all-speed mechanical one. This controls only the idling and maximum speeds; at intermediate speeds the metering valve is under the direct control of the accelerator. This makes the engine response more immediate, eliminates any idle pedal movement and can improve fuel consumption.

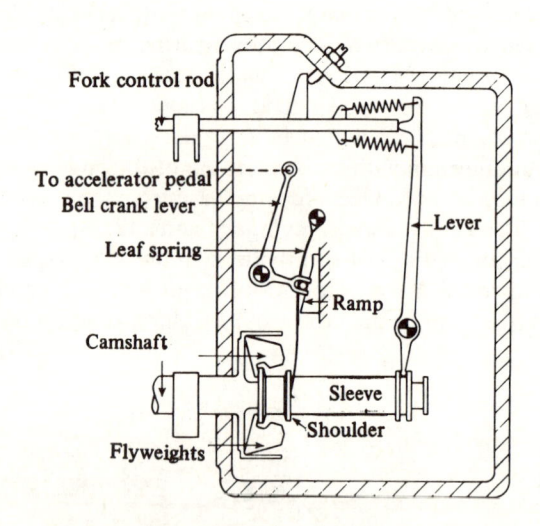

Figure 12.14   Leaf-spring governor: schematic

12.16   Cold starting

When cold starting a compression-ignition engine, heat loss from the compressed air may lower its temperature below that needed for self-ignition of the fuel. The problem is more acute in indirect-injection engines with a larger cold surface area, and will always be exacerbated by a slow cranking speed.

Indirect-injection engines normally require battery-supplied heater plugs in the swirl chambers. The resistance filament reaches up to 1000°C and provides both a heat source and an incandescent surface for the injected fuel.

For direct-injection engines an inlet manifold heater can be used to vaporize and ignite a small quantity of fuel. In a typical device, the

The compression-ignition engine

expansion of an electrical heater coil admits and meters the fuel, which is then fired by an igniter coil within a surrounding perforated shield.

Excess fuel delivery for starting can be provided on both in-line and distributor-type pumps by an override of the maximum fuel delivery stops. Because of the possibility of a driver causing HC smoke emission whilst attempting to increase power in this way, legislation requires the device to automatically latch out after starting and be unavailable while driving. The provision is technically simple on the in-line pump where the control rod is allowed extra movement. Regulation of the limiting travel of the plungers in the distributor pump demands more complex devices, operating automatically at cranking speed either from internal fuel pressure or from the position of the throttle arm.

Starter fluids

Ether-based fluids have a low self-ignition temperature and readily ignite under CI compression. They may be introduced from an aerosol spray into the air cleaner or from an installed system, manually operated. Improper use of these fluids can lead to severe loading and potential mechanical failure.

12.17 Injectors

The function of the injection nozzle is to deliver fuel so that it can be completely burnt in the cylinder. This entails no leakage before injection, correct shape of the spray and atomization during injection, and no dribble after injection (Figure 12.15).

The nozzle consists of a valve and a body, the parts being fitted with extreme accuracy and non-interchangeable. The fuel feeds to a gallery via drillings in the body, and increasing pressure from the fuel pump raises the valve against a spring contained in the nozzle holder (Figure 12.16).

The profile of the pintle needle and nozzle can be designed to provide the spray characteristics required and to limit the initial fuel delivery, minimizing the delay period and subsequent diesel knock (Figure 12.17). The pintaux (pintle/auxiliary) nozzle has a further modification where at small lift an auxiliary spray promotes easy starting (Figure 12.18).

The injection pressure is controlled by the spring and can vary from some 10 MPa for a pintle-type nozzle used with indirect injection to 30 MPa with single- or multi-hole nozzles.

Injectors should be removed for testing every 20 000 to 25 000 km. A faulty injector can cause misfiring, knocking, smoky exhaust, loss of

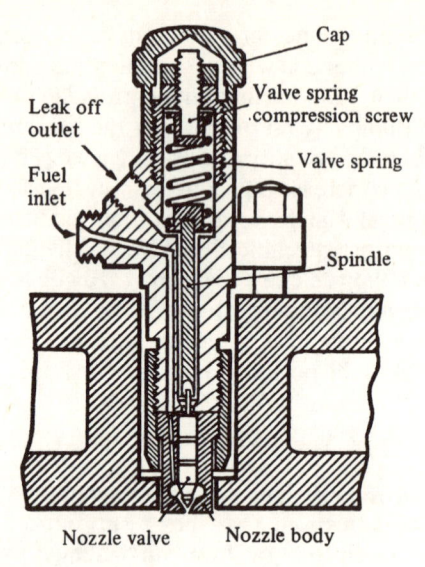

Figure 12.15 Fuel injector

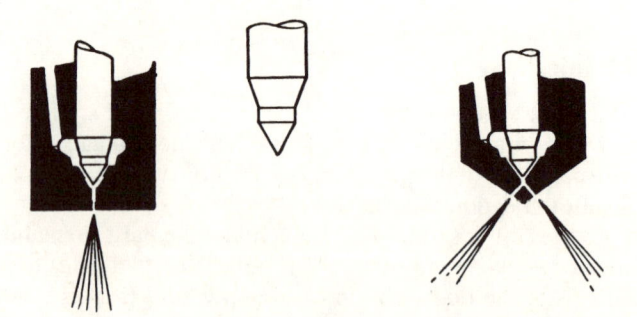

Figure 12.16 Single- and multi-hole nozzles and valve: direct injection

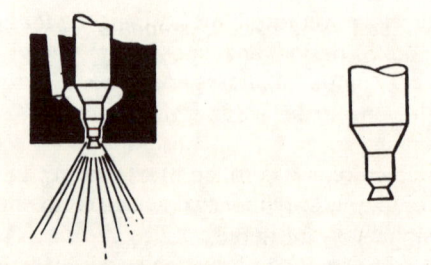

Figure 12.17 Pintle nozzle and valve: indirect injection

The compression-ignition engine

Auxiliary hole

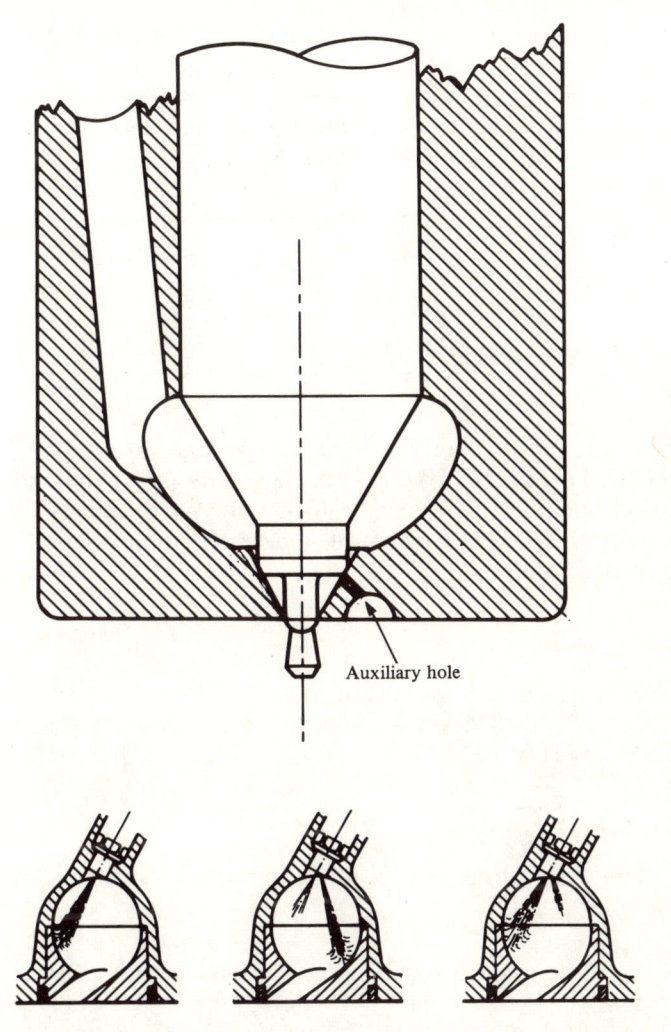

Figure 12.18 Pintaux nozzle. Left: standard pintle. Centre: pintaux, starting. Right: pintaux, running

power, overheating and high fuel consumption. It can be located by running the engine at some 1000 rev/min and slackening the union nut at the end of each delivery pipe in turn. Cutting-out the faulty injector will have little or no effect on the speed.

To test an injector on the vehicle, it may be removed and refitted outwards, and turned away from the operator. Then, with the other injector unions slackened to prevent starting, rotation of the engine will allow examination of the spray for 'solid' streaks of fuel, incorrect shape or choked nozzle holes. A replacement injector should be fitted if these are evident. Reconditioning the nozzles and resetting the injection pressure requires appropriate servicing equipment.

Great care must be taken to avoid the spray from any injection nozzle, as it will easily penetrate the flesh. The fuel oil itself can be harmful if frequently handled, and a neutral oil of similar viscosity is used during the overhaul and testing of the injection equipment.

12.18 Maintenance

Fuel

Fuel cleanliness is of the highest importance. Main fuel filters are replaced at 20 000–25 000 km intervals. Lift-pump filters and air filters are cleaned or replaced every 10 000 km. Water and sediment bowls require frequent inspection; in some cases warning light sensors are provided.

Lubrication

Heavy-duty diesel-type detergent oils are required and should be changed every 5000 km. Full-flow filters are usually replaced at 5000–10 000 km intervals.

Cooling

Overheating can damage the injection-nozzle valves and seats.

Starting

Slow cranking exacerbates starting difficulties. Poor compression, due to defective valve, rings or cylinder condition, will cause difficult starting, particularly in cold weather. Where a cold-starting aid introduces fuel oil or ether-based fluids into the inlet manifold, incorrect use or leakage can cause engine damage.

Smoke

Exhaust smoke signifying imperfect combustion may be due to:

(a) Choked air filter
(b) Faulty injection nozzles or delivery valves
(c) Incorrect fuel-injection pump timing or maximum delivery setting

(d) Defective cold-starting aid
(e) Lubricating oil – from worn rings, bore or defective valve-stem oil seals.

12.19 The turbocharger

Unlike mechanically driven superchargers, the turbocharger utilizes waste heat energy from the exhaust gas and can improve both thermal efficiency and power output. The turbocharger narrows the performance gap between equal-capacity compression and spark ignition engines and, by reducing the delay period, improves smooth running.

The exhaust gas drives a radial-flow turbine with a centrifugal compressor on the other end of the same shaft (Figure 12.19). Maximum speeds are in the range 80 000–120 000 rev/min and maximum exhaust temperatures 650–900°C. The plain bearings are lubricated from the engine system.

Water after-cooling will remove compression heat from the air and provide a more dense cylinder charge as well as reducing peak cylinder temperatures.

The boost pressure varies with engine speed, and rotational inertia causes some turbine lag during acceleration. A diaphragm control, operated by inlet manifold pressure, can be fitted to both in-line and distributor-type pumps to modify the maximum fuel delivery with boost pressure, so obtaining maximum performance whilst avoiding smoke emission.

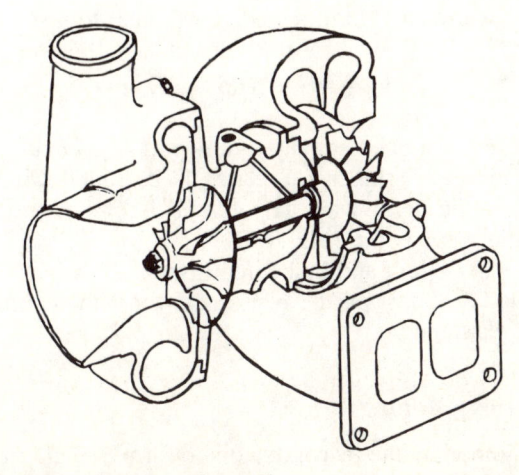

Figure 12.19 Turbocharger: schematic

Chapter 13

The electrical system

An electrical circuit in which a dynamo or battery produces an electrical pressure of say 12 volts, and causes a current (a movement of electrons) to flow around the conductor at a rate of 3 amperes, may be compared to a water pump creating a hydraulic pressure and producing a flow of water around a pipeline.

13.1 Ohm's law

Ohm's law states that the current in a circuit varies in direct proportion to the pressure (voltage) and in inverse proportion to the resistance of the circuit:

$$I = \frac{V}{R}$$

where I = current in amperes (A), V = pressure or potential difference in volts (V), and R = resistance in ohms (Ω).

Ohm's law can also be expressed in two other forms:

$$V = IR \quad \text{and} \quad R = \frac{V}{I}$$

The law can be applied equally to any part of a circuit as well as the whole circuit, and in this case it relates the current flowing between any two points in the circuit with the potential difference and the resistance between these two points (Figure 13.1).

In a simple battery and bulb circuit, the bulb, which is the resistive load, provides almost all the resistance whilst the conductors have a minimum resistance.

13.2 Series grouping

Resistors joined in line by conductors to form a single circuit, with the same current flowing through each of them, are said to be connected in

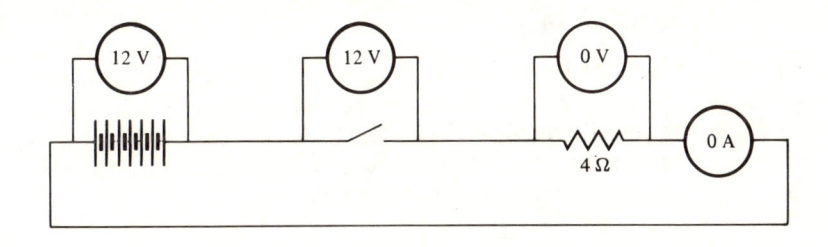

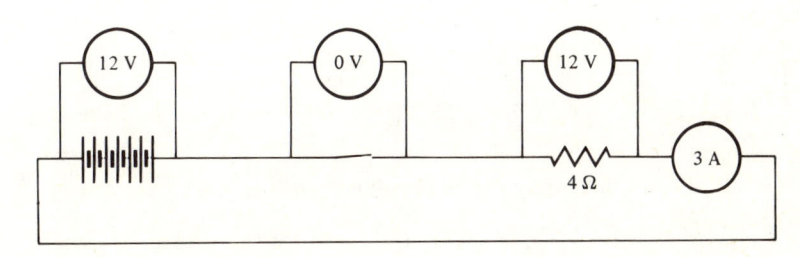

Figure 13.1 Series circuit: off and on

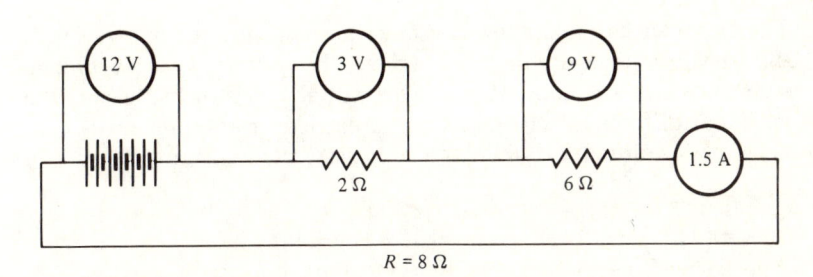

Figure 13.2 Series circuit

series (Figure 13.2). The total resistance of a series circuit is the sum of the individual resistances. This will increase and the current will decrease as more resistors are added in series:

$$R = r_1 + r_2 + r_3 + \ldots$$

where R = total resistance (Ω) and r_1, r_2, r_3 = individual resistances (Ω).

13.3 Parallel grouping

Resistors joined by conductors so as to form a number of alternative paths for the current are said to be connected in *parallel* (Figure 13.3). The total current in a parallel circuit is the sum of the currents in the individual branches. This will increase and the total resistance will decrease as more resistors are added in parallel:

$$I = i_1 + i_2 + i_3 + \ldots$$

where I = total current (A) and i_1, i_2, i_3 = currents in parallel paths (A). As $I = V/R$ applies to the whole circuit and to the individual branches

$$\frac{V}{R} = \frac{V}{r_1} + \frac{V}{r_2} + \frac{V}{r_3} + \ldots$$

The total resistance in a parallel circuit is therefore given by

$$\frac{1}{R} = \frac{1}{r_1} + \frac{1}{r_2} + \frac{1}{r_3} + \ldots$$

13.4 Ammeter and voltmeter

The ammeter has a very low resistance and is connected in series with the resistor to measure the current flow. The voltmeter has a very high resistance and is connected in parallel with the resistor to measure the potential difference across it. Care is needed; if an ammeter is connected as a voltmeter it will pass a large destructive current.

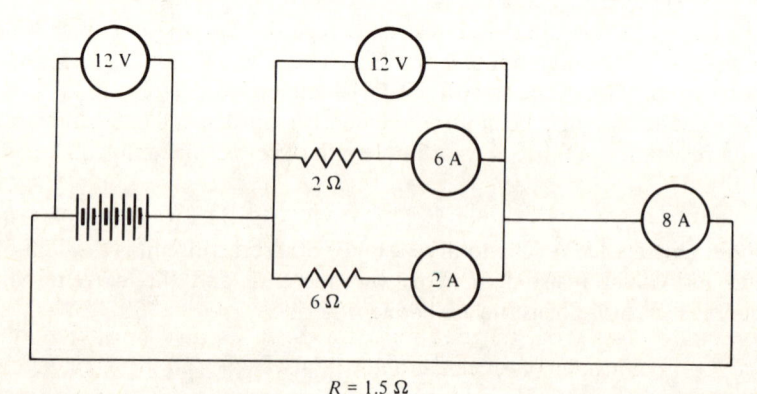

$R = 1.5\ \Omega$

Figure 13.3 Parallel circuit

13.5 Electrical power

The electrical unit of power is the watt (W). Electrical power is given by

$$P = VI$$

where P = power in watts (W), V = potential difference or voltage (V), and I = current (A). In addition:

$$P = I^2 R$$

and

$$Q = I^2 Rt \quad \text{or} \quad Q = IVt$$

where R = resistance (Ω), t = time (s), and Q = heat generated in joules (J); 1 joule = 1 watt second.

13.6 Conductors and cables

All pure metals are good conductors. Silver is the best; it offers the least resistance to the flow of electricity. Copper is almost as good and, being much cheaper, is very widely employed.

The *resistance* of a conductor not only depends on the material but also varies in direct proportion to its length and inversely with the cross-sectional area (CSA):

$$R = \rho \frac{l}{a}$$

where R = resistance (Ω), l = length (m), ρ = resistivity (Ωm), and a = cross-sectional area (m^2).

The *resistivity* of a material can be regarded as the resistance of a length of 1 m of material of 1 m^2 CSA. Thus the units of resistivity are Ωm^2/m or Ωm. The resistivity of commercial copper at 20°C is 0.018 $\mu\Omega$m; this may be more conveniently regarded as 0.018 Ωmm^2/m, or a resistance of 0.018 Ω for a 1 m length of copper of 1 mm^2 CSA (see Table 13.1).

The resistance of the usual conductors increases with rise in temperature – copper and aluminium by about 0.004 Ω per 1°C. Carbon is exceptional for its decrease in resistance with increase in temperature.

To avoid an excessive voltage drop and overheating, cables of adequate cross-sectional area must be used. To give flexibility, the cables are normally of several strands. For example, a 14/0.30 cable has 14 strands of wire each 0.30 mm diameter, giving a CSA of approximately 1.0 mm^2, and is suitable for currents up to 8.75 A, as for side, tail, indicator and stop lamps and ignition wiring. Other generally used cable sizes are 28/0.3 for headlamps, 44/0.3 and 65/0.3 for charging circuits – with maximum continuous currents of 17.5 A, 27.5 A and 35 A respectively. All these cables produce a voltage drop of about 0.15 volts per metre (V/m) at the maximum rated currents. The total voltage drop in any circuit when fully loaded must not exceed 10 per cent of the applied voltage, and preferably should be under 5 per cent.

Table 13.1

Substance	Resistivity (Ω mm^2/m at 20°C)
Silver	0.016
Copper (commercial)	0.018
Aluminium	0.029
Zinc	0.059
Nickel	0.070
Iron	0.098

13.7 Circuits

In general, the electrical system comprises a number of simple circuits in parallel. Each consists of a feed wire, a switch, a switch wire, the component and an earth wire if required. The earth return circuit is then completed through the chassis of the vehicle to the earthed terminal of the battery. Where flammable materials are carried, any shorting of a live wire on to metalwork could be hazardous, and such vehicles use insulated return wires; these are also necessary with glassfibre bodywork.

The separate cables are braided into a protective harness to give mechanical strength. Identification of the individual leads is provided by colour coding and their position in the loom. A base colour with thin coloured tracers is used on the insulation, but while these colours are generally standardized in a particular country there is no international agreement. Snap-in bullet and flat-pin connectors are widely used; they may be grouped into junctions and multiway connections for coupling sublooms into the main system. In the instrument panel separate wiring can be economically replaced by the printed circuit board – an insulating board with bonded copper foil. Overloads must be avoided as they can easily fuse the foil.

13.8 Headlamps

A tungsten filament, heated to some 2500°C in an inert gas and accurately located at the focus of a parabolic reflector, whose beam is modified by the cover-glass lens, provides the main illumination. A second filament, arranged so that the rays are reflected downwards, produces the dipped beam. The semi-sealed light unit uses a pre-focused bulb located by its flange in the reflector-lens assembly. In the sealed-beam system the filaments are an integral part of the gas-filled reflector-lens unit.

In the tungsten-halogen bulb the inert nitrogen is replaced by a gas containing one of the halogen group – usually iodine or bromine – at a higher pressure in a quartz envelope. This allows the filament to operate at a temperature approaching 3000°C and recycles the evaporated tungsten. Light output and filament life is increased, whilst blackening of the bulb from deposited tungsten is avoided. The quartz glass should not be handled, and if inadvertently touched should be cleaned with methylated spirit.

13.9 Fuses

The fuse is a safety device in a circuit to prevent overloading. It consists of a suitably thin wire or strip which will melt (fuse) if an excessive current passes through it, and so break the circuit. Ceramic fuses are marked with the maximum current they can carry, glass cartridge fuses with their fusing value – about twice the continuous current.

The different circuits have separate fuses to prevent a failure affecting several components. The ignition, however, is not usually fitted with a fuse, since a sudden failure might be dangerous. A small number of fusible links – extra high-rate fuses – are often installed near the battery to prevent a fire hazard from a massive short-circuit – as during an accident.

If a fuse fails, the cause of the overload should be located and rectified before the fuse is replaced.

13.10 Wiring diagrams

Wiring diagrams from different manufacturers can range from a simplified presentation of the electrical connections to an elaborate illustration of the physical layout of the components, loom details, internal connections etc. Whatever system is employed it should be possible to identify any required circuit and connection using the symbols, number and letter codes – which are not standardized.

Chapter 14
The ignition system

Gases are very poor conductors of electricity; only a small gap, as in a switch, is needed to open a low-voltage circuit. To ionize the compressed gas between the sparking-plug points and produce a spark requires an electrical pressure of 5–20 kV depending on the conditions. To initiate combustion a spark with a minimum duration of about 300 μs and with some 0.03 J of energy is necessary.

14.1 Fundamental laws

When a current flows in a circuit it creates a *magnetic field* around the conductor, and this field will, in effect, move outward and inward as the current increases and decreases. The magnetic field can be strengthened by coiling the wire into a *solenoid*. If the solenoid is fitted on to a soft-iron core, preferably with insulated laminations to reduce the induction of unwanted eddy currents, an *electromagnet* having a powerful magnetic field can be produced (Figure 14.1).

When a magnetic field moves across a conductor an *electromotive force* (EMF) is induced and is in proportion to the rate of movement. If a complete circuit exists a current will flow (Figure 14.2).

Self-induction occurs when the current in an inductive circuit changes and the magnetic field cuts the conductors. This induced electromotive force opposes the change in current, restricting it if the current is increasing and enhancing it if the current is decreasing.

Mutual inductance takes place when a current in one winding induces an electromotive force in another winding in the same magnetic circuit. The induced EMF is in proportion to the number of turns in each winding.

14.2 Battery-coil ignition

The primary or low-tension (LT) circuit on a vehicle contains the battery, ignition switch, coil primary winding and contact breaker (CB)

The ignition system

(a) Magnetic field surrounding magnets

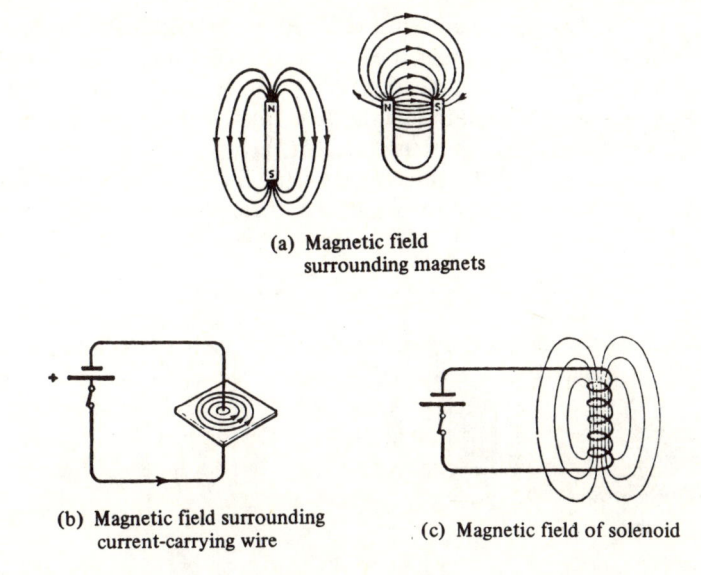

(b) Magnetic field surrounding current-carrying wire

(c) Magnetic field of solenoid

Figure 14.1 Production of a magnetic field

(a) Current induced when magnetic field cuts a conductor

(b) Induction of LT current

(c) Induction of HT current

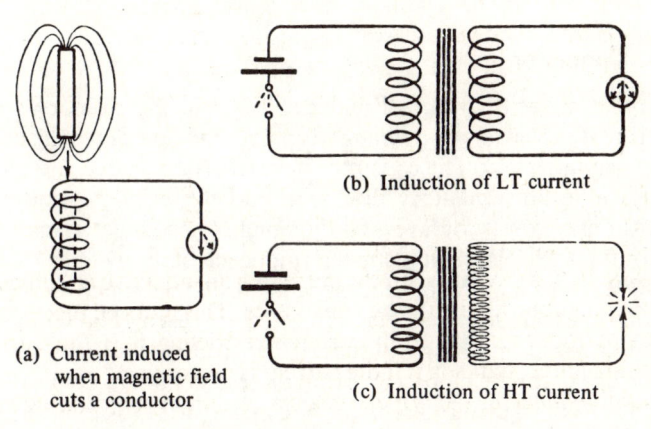

Figure 14.2 Electromagnetic induction

with capacitor in parallel. These are linked by wire connections or earth return (Figure 14.3).

With the ignition switch and contact-breaker points closed a current of 2–4 A flows around the 300 or so turns of the coil primary; the primary is the outer of the two windings so that heat may more easily dissipate. A magnetic field is produced, strengthened by the laminated core and soft-iron sheath surrounding the windings. When the CB points are separated by the rotating cam, 200–300 V are self-induced in the primary winding. Mutual induction then creates up to 20 000 V in the secondary winding – some 30 000 turns of fine wire.

The high-tension (HT) current passes to the rotor arm, across the small air gap to the distributor segment and then to the sparking plug.

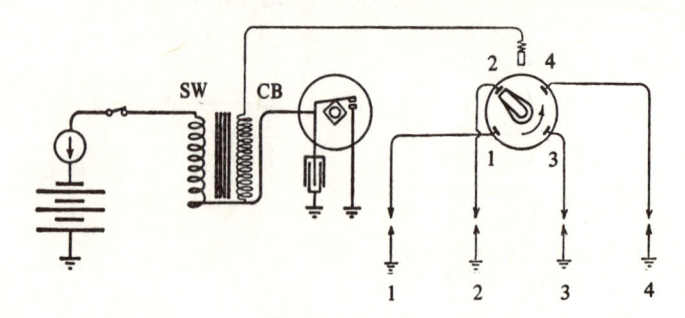

Figure 14.3 Battery-coil ignition

14.3 Capacitor

Without a capacitor, the self-induced EMF in the primary winding would create an arc at the opening CB points and slow down the collapse of the magnetic field. The capacitor stores this destructive energy, then releases it in an oscillatory discharge into the primary windings. A capacitor may consist of layers of aluminium foil connected to opposite sides of the CB points but insulated from each other by mica or waxed paper. More usually metallized paper – insulating paper on to which very thin metallic foil has been deposited – is used. The coil ignition condenser has a capacitance of some 0.2 microfarad (μF).

If the capacitor develops a short circuit (e.g. failure of the insulation) the contact points cannot break the circuit and no HT is induced. If the capacitor has an open circuit, arcing at the CB points will occur and the HT spark will be weak and erratic.

14.4 The distributor

Since all cylinders must fire in two revolutions of any four-stroke engine, the four-cylinder distributor has a four-lobe cam (and the six-cylinder unit a six-lobe cam) driven at half crankshaft speed by skew gear from the camshaft, often an extension of the oil-pump drive, or directly from the end of an overhead camshaft.

The rotor arm mounted above but insulated from the cam receives the HT current from the coil through a contact in the distributor cap and a carbon brush and distributes it to the segments which are connected to the plugs in the sequence required by the firing order.

14.5 Polarity

The battery negative terminal is earthed to the chassis. The coil terminals are marked + and −, the negative terminal being connected to the contact breaker for the negative earth system.

The coil is designed to produce negative HT polarity at the hotter central electrode of the sparking plug. If the cooler earthed electrode were negative, a higher voltage would be required to produce the spark.

14.6 Ballasted ignition circuit

The effectiveness of charging systems permits batteries of reduced capacity to be fitted. The ballasted ignition circuit is used to avoid starting difficulties consequent on the reduced battery voltage available at the coil when the starter motor is operating.

The ignition coil has a lower-resistance primary winding and a satisfactory output on the reduced voltage, some 9 V. It is usually supplied from an extra terminal on the starter solenoid switch, energized during starter operation. At other times the coil receives a similar voltage through the ignition switch and a ballast resistor, which may be a mounted unit or a resistive lead.

14.7 Advance and retard

To obtain the maximum power and economy it is necessary to advance the spark with increase in engine speed since, although the rate of flame travel increases nearly as quickly as the speed, the delay (in time) between the spark and the start of combustion is constant and so increases in terms of crank-angle rotation as the speed increases.

An automatic mechanism acting as a flexible coupling is incorporated in the distributor drive. A pair of rotating weights, operated by centrifugal force against springs, twist the cam in its direction of rotation relative to the drive shaft, so opening the CB points earlier at higher speeds. By suitable arrangements of the springs, weights and stops the advance characteristics can be made to conform approximately to the optimum engine requirements (Figure 14.4).

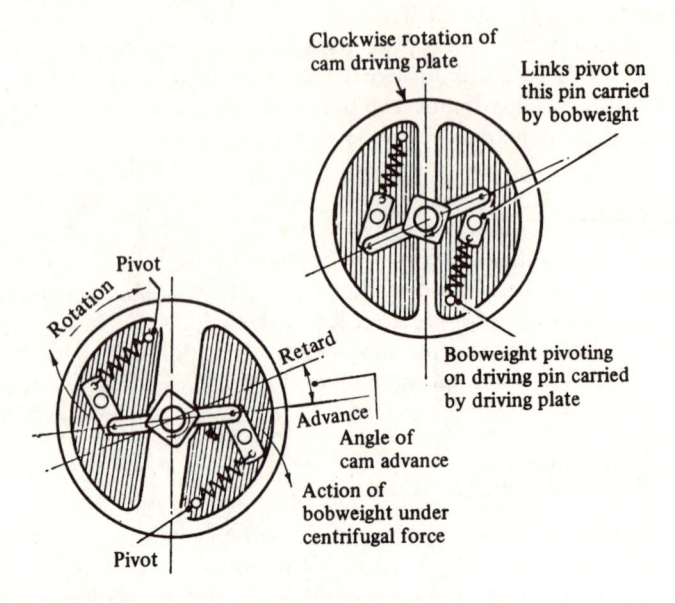

Figure 14.4 Centrifugal ignition-timing control

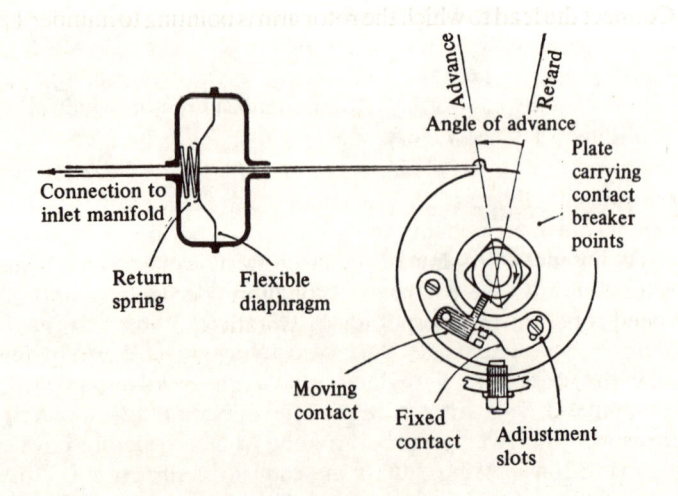

Figure 14.5 Vacuum ignition-timing control

An additional vacuum control is usually fitted which consists of a spring-loaded diaphragm connected to a rotatable plate carrying the CB points (Figure 14.5). The diaphragm is subject to the depression in the induction manifold via a small drilling located slightly on the air-intake side of the closed throttle valve. At light-load part-throttle conditions, where a slower-burning economy mixture is provided, the considerable depression in the manifold advances the ignition. With open-throttle heavy-load conditions and also at idling speed, the depression is reduced and the return-spring acting on the diaphragm retards the spark.

By means of these devices the ignition timing can be related to both the load and the speed of the engine. An additional manual adjustment is sometimes provided to allow slight variations of the static timing without necessitating rotation of the distributor body.

14.8 Timing the ignition

(a) Turn the crankshaft until piston number 1 is at TDC of the compression stroke, i.e. both valves are closed and the static timing marks on the pulley or flywheel are in line.
(b) Set the manual adjustment, if fitted, to the mid position.
(c) Turn the distributor body, in the opposite direction to cam rotation, until the CB points are just opening.
(d) Connect the lead to which the rotor arm is pointing to number 1 plug, and connect the other segments around the distributor cap to the plugs in the sequence of the firing order. The firing order and the compression stroke can be determined by noting the valve action.
(e) With the engine running and a stroboscopic lamp directed at whitened timing marks, check the centrifugal advance control with the vacuum pipe disconnected, then the combined advance devices, against the maker's data.
(f) In the absence of equipment, reliance must be placed on a road test. Excessive advance produces detonation (particularly under low-speed, open-throttle conditions), vibration, a lack of power and abnormal bearing loads. Over-retarded ignition timing causes a lack of power, with a flat exhaust note, poor economy and the risk of overheating.

14.9 Sparking plugs

The central electrode, usually of a nickel steel, is embedded in an aluminium-oxide insulator. The plug body, occasionally detachable for cleaning for special applications, with 18, 14, 12 or 10 mm thread diameter and varying reach, makes a gas-tight joint around the insulator and carries the earthed electrode – which alone must be bent during gap setting (Figure 14.6).

Correct reach is important, both for ignition efficiency and to avoid future problems where a shorter reach can allow carbon to choke the unused inner threads.

Plugs are also classified according to their resistance to heat or oil. A grade is required which will run hot enough to burn off oil and soot and so prevent misfiring, but not hot enough for the central electrode to become incandescent and cause pre-ignition.

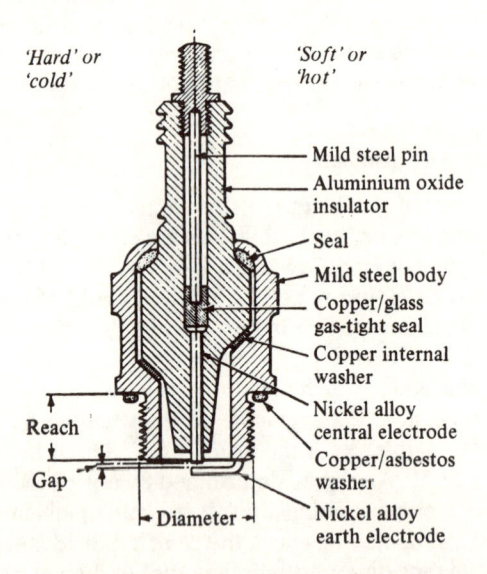

Figure 14.6 Comparison of sparking plug types: hard and soft

If the correct heat-grade plug is fitted, the insulator appearance is a rough guide to the operating conditions. A light tan to a greyish brown is normal; overheating, possibly from a weak mixture, produces a white glazed appearance, whilst cool running with a rich mixture results in sooty deposits.

14.10 Maintenance

Before sparking-plug removal all accumulated dirt must be cleaned from the plug recesses. Plugs should be cleaned, reset and tested every 10 000 km. Testing must be carried out under compressed air; a faulty plug might not be detected at atmospheric pressure. Replacement rather than servicing is usual. Correct torque when refitting is important, whether sealing is by an attached gasket or by a tapered seat.

HT leads have a graphite-impregnated nylon core of specified resistivity to suppress radio interference. Replacements when required should be of the original type and length; alternative resistivities are available.

The CB points should be checked every 10 000 km and replaced if burned or pitted; cleaning is generally impractical with current design. The CB gap should be set, using a dwell meter, to the specified angle – some 50° for four-cylinder and 40° for six-cylinder units. Dwell angle is the angle of distributor-cam rotation between the CB points closing and opening again, when the coil magnetic field is being produced by the primary current. When a (clean) feeler gauge is used to set a specified gap – 0.35–0.6 mm according to type – a pit-and-pip condition of the contact faces, due to metal transfer, can result in a faulty clearance. An increased gap will advance the ignition timing and reduce the dwell angle.

When lubricating parts of the distributor, care is necessary to avoid any oil reaching the contact points and causing pitting. The insulating surfaces of the distributor cap, rotor arm, HT leads and plugs should be cleaned to prevent HT tracking (current leakage).

14.11 Electronic induction discharge ignition

There are two basic electronic ignition systems. The first follows orthodox practice already described, where interruption of the primary current induces the secondary voltage. However, by using transistors – very high-speed solid-state switches activated by small trigger currents – to control the primary circuit, two advantages arise. If (unusually) contact points are retained, they handle only the small trigger current – under 0.5 A – and metal transfer is much reduced. Usually some form of transducer replaces the contact breaker. A rotor blade can interrupt a light or infrared beam, or the magnetic balance of a suitable pick-up can be disturbed by revolving magnets, ferrite rods or vanes. In each case the transducer signal is used to trigger the transistor system.

The second advantage of using transistors is that the primary current can be almost doubled, allowing the use of coils with fewer primary turns and a higher primary/secondary ratio (up to 1:400 turns). The lower-inductance coil allows the primary current to reach its maximum in a shorter time, and enables the HT voltage to be maintained at higher engine speeds.

14.12 Capacitor discharge ignition

A capacitor, charged at about 400 V, is discharged through the coil primary by a thyristor – a development of the transistor, which remains conductive after receiving a pulse from the transducer (or in a few cases the contact-breaker points). To charge the capacitor, battery direct current is first converted to alternating current, transformed to the higher voltage, and then rectified to direct current again. A low-inductance coil is used with the system.

Capacitor discharge ignition (CDI) has an extremely rapid current rise time, and it is claimed that this high-energy system will operate satisfactorily under adverse conditions of plug fouling and leakage. Some consider, however, that the short spark duration can produce unreliable ignition in weak mixtures.

The electrical energy of CDI systems can be dangerous.

14.13 Electronic ignition control

The ignition electronic control unit (ECU) is a microprocessor with a memory programmed to the basic ignition requirements of the engine. It receives signals from a chosen number of sensors and computes them to direct the timing, and if desired the energy, of the spark produced by the selected HT system – induction or capacity discharge.

If the flywheel carries suitable triggers, transducers in proximity can provide the ignition datum point, e.g. TDC, and signal the engine speed. Engine load can be obtained, for example, from an inlet manifold pressure unit, whilst knock sensors in the cylinder block can assess incipient detonation and activate a chosen timing retardation.

Additional relevant signals can be incorporated as desired, and the unit may be linked with a carburettor or fuel-injection ECU to provide so-called electronic engine management.

Chapter 15
The charging and starting systems

15.1 The battery

A storage battery is necessary for starting, lighting and the electrical accessories in addition to the coil ignition system.

When a current is passed through a cell consisting of two lead plates immersed in dilute sulphuric acid, a chemical action takes place converting the surface of the anode – the plate connected to the positive supply – into lead peroxide and the cathode surface into spongy lead. This action is termed *electrolysis*, and the acid solution is the *electrolyte* (Figure 15.1). If the charging leads are now disconnected there will be a voltage between the plates, and when connected by an external circuit a discharging current will flow in the opposite direction. As the cell discharges, both anode and cathode surfaces become converted into lead sulphate, the acid solution becomes weaker, with a lower specific gravity, and the voltage falls. Recharging will convert the anode to lead peroxide and the cathode to spongy lead, and restore the acid strength.

The motor vehicle battery of the lead-acid type consists of a robust insulated case with a number of cells each containing alternate positive

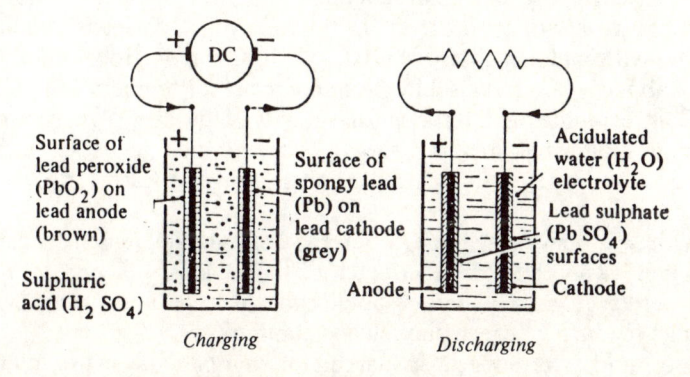

Charging: Surface of lead peroxide (PbO_2) on lead anode (brown); Surface of spongy lead (Pb) on lead cathode (grey); Sulphuric acid (H_2SO_4)

Discharging: Acidulated water (H_2O) electrolyte; Lead sulphate ($PbSO_4$) surfaces; Anode; Cathode

Figure 15.1 *Electrolysis in secondary cell*

and negative plates with porous insulators between them – each group of the same polarity being connected to an external terminal. The voltage of each cell when fully charged is about 2.5 volts and the specific gravity about 1.280 (for temperate climates). As discharging commences, the voltage drops to 2.2 volts and, when it falls to 1.8 volts and the specific gravity to about 1.150, the cell must be recharged. The cells are connected in series to form a 6-volt or 12-volt battery.

The *capacity* of the battery – the current it can maintain for a certain period – is primarily determined by the size and number of plates in each cell. However, capacity also depends on the rate of discharge, and in stating capacity it is usual to assume a 10-hour rate. Hence a '75 ampere hour' battery should provide 7.5 A for 10 hours without the voltage falling below 1.8 V; it might provide 4.5 A for 20 hours but 15 A for under 4 hours.

Maintenance

For safety, the battery must be firmly secured in the vehicle. Petroleum jelly inhibits terminal corrosion, which – if present – will readily dissolve in hot water. The battery should always be disconnected before any work on the vehicle which might produce a short circuit to an electrical connection.

Periodical topping-up with distilled water is necessary to maintain the electrolyte above the plates. The state of charge can be determined using either a hydrometer or a heavy-discharge instrument. The latter is a voltmeter with a low resistance in parallel to allow a heavy current to pass whilst testing (about 150 A for batteries under 75 Ah and 300 A for 75 Ah and larger capacity).

For accuracy, hydrometer readings must not be taken following topping-up or with an incorrect electrolyte level. A correcting addition to the hydrometer readings of 0.007 per 10°C for electrolyte temperatures above 15°C, and a subtraction to correct for temperatures below 15°C, is recommended. If the specific gravity of the electrolyte is suspect through spilling etc., readings should be taken when the battery is fully charged, i.e. when 'gassing' freely and showing no further rise of the SG.

With the heavy-discharge tester readings should remain constant between 1.2 and 1.5 V for each cell for 5–10 seconds.

A battery in good condition should show little variation in the SG or voltage readings between one cell and another.

Too rapid or prolonged discharging or charging may cause internal defects. If the battery is to be stored it must be periodically charged,

The charging and starting systems

otherwise the lead sulphate formed will not reconvert to lead peroxide and lead, and the plates may buckle and shed the active material.

During charging, hydrogen and oxygen are evolved, and a spark or flame near the vents may well cause a dangerous explosion.

It is important, and particularly with AC equipment, to ensure the correct 'earth' polarity for the battery connections; British practice since 1966 conforms to the international standardization of negative battery earthing.

15.2 AC and DC generators

When a magnetic field is cut by a moving conductor, a current is induced in the conductor in a direction indicated by Fleming's right-hand rule (Figure 15.2). If the conductor is in the form of a revolving coil with its

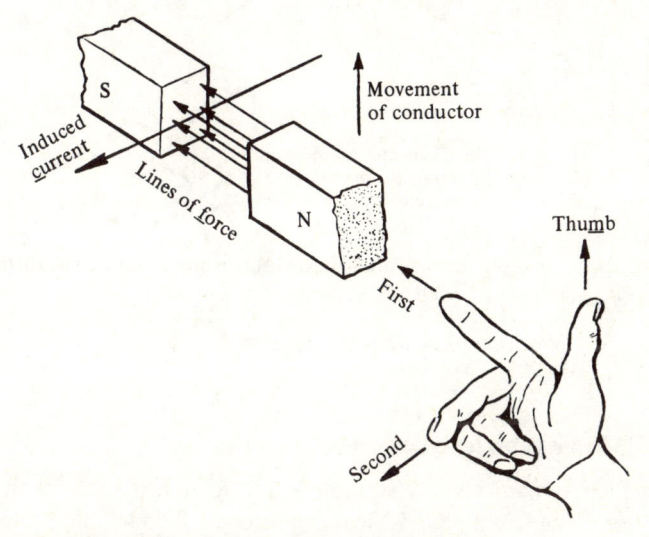

Figure 15.2 Fleming's right-hand rule for generators

ends connected to two slip rings on which rest stationary brushes connected to an external circuit, then an alternating current (AC) will be induced. This current flows first in one direction, diminishes to zero, increases to a maximum in the opposite direction, decreases to zero, and repeats the cycle as the coil makes a further revolution (Figure 15.3).

By attaching the ends of the revolving coil to a commutator – a drum formed of copper strips with mica insulation – a direct current (DC), flowing in one direction only, can be collected by the brushes (Figure 15.4).

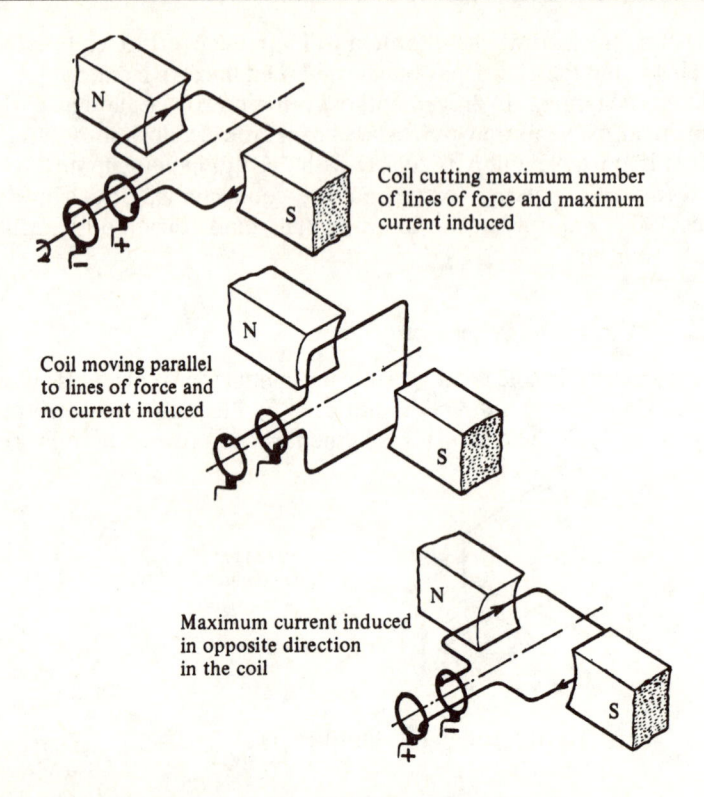

Figure 15.3 Production of alternating current

15.3 Motor vehicle dynamos

The vehicle dynamo is a DC machine having an armature, with many coils, revolving in a magnetic field produced by field windings mounted on pole pieces attached to the carcase (Figure 15.5). The ends of the armature coils are connected to the separate bars of the commutator from where the current is collected by two brushes. A proportion of this current is by-passed or shunted around the field windings, and since the dynamo voltage would otherwise rise in proportion to its speed of revolution, an automatic regulator is fitted in this field circuit.

Current/voltage control

The control unit contains an electromagnet with the soft-iron core carrying a voltage winding – considerable turns of fine wire shunted

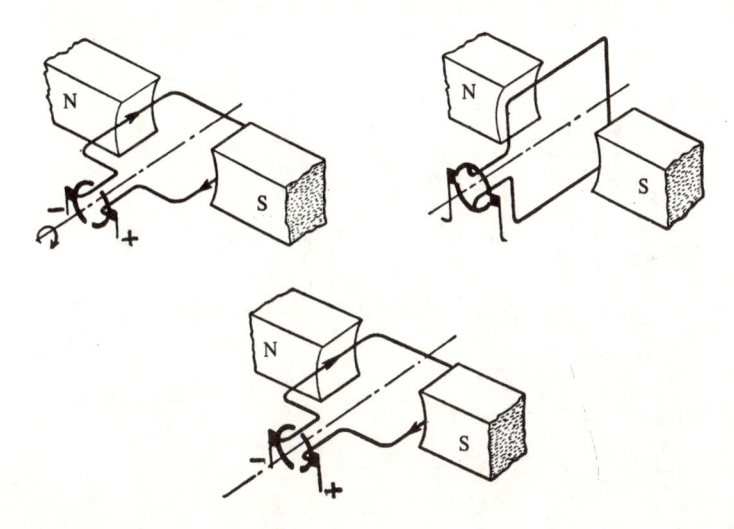

Figure 15.4 Production of direct current

Figure 15.5 Arrangement of pole pieces

across the dynamo output. A spring-loaded keeper or armature operates contacts which, when closed by the spring, by-pass a resistance in the field circuit (Figure 15.6).

The voltage of the dynamo rises until the field of the electromagnet becomes sufficiently powerful to overcome the spring tension and open the regulator contacts. This brings the resistance into the field circuit, reducing the field current and weakening the magnetic field cut by the revolving armature coils. The dynamo voltage falls and the regulator contacts close, short-circuiting the field resistance; the voltage rises again and a vibrating action of the contacts results.

By maintaining a constant dynamo voltage irrespective of speed, the charging current delivered to the battery is suited to its state of charge.

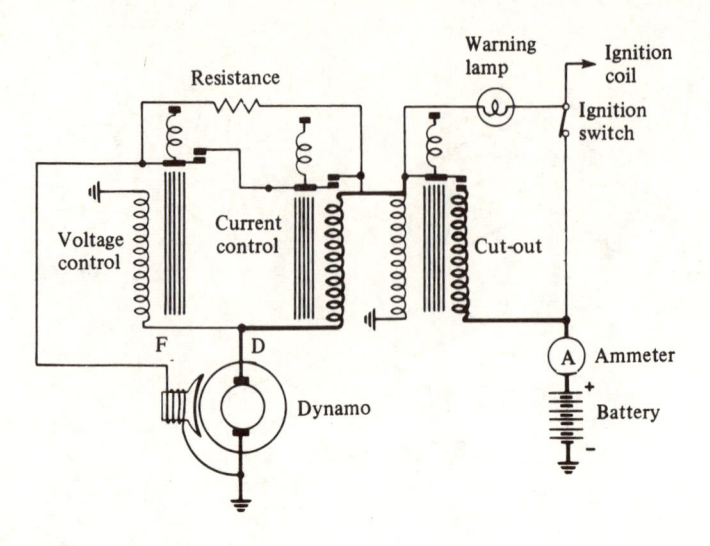

Figure 15.6 Dynamo control circuits

A fully charged battery having a high voltage will receive only a small trickle charge, whilst the low voltage of a discharged battery will cause it to receive a large current, since the current flowing depends upon the difference between dynamo and battery voltage.

With only voltage regulation excessive current could be produced, as with a completely discharged battery. A current regulator is therefore also fitted. When the current reaches the rated maximum for the dynamo, the regulator contacts are set vibrating by the current coil, introducing the field resistance and controlling the output.

The cut-out

At slow speeds the voltage produced by the dynamo is lower than the battery voltage and the battery is prevented from discharging through the dynamo by the cut-out, an automatic switch incorporating an electromagnet with two windings.

The ignition warning lamp is shunted across the cut-out contacts and is therefore short-circuited by the closed contacts when charging commences.

The voltage winding, having many turns of fine wire, is shunted across the dynamo output and, when the dynamo voltage is sufficiently high, the magnetic field attracts an armature against spring tension and closes the cut-out contacts. The closed contacts allow the charging current to flow around the current coil which is in series with the ammeter and battery.

As long as the charging current flows, the magnetic fields of the two coils assist each other. However, when the dynamo voltage falls below the battery voltage, a reverse current flows through the current coil and the two magnetic fields oppose each other; the armature is released and the charging circuit is broken by the cut-out contacts.

15.4 Motor vehicle alternators

Semiconductors

Three types of solid-state semiconductor are used with the alternator charging system. The *diode*, a one-way valve, allows current to pass in one direction only. Diodes are used to rectify the alternating current, the function of the commutator in a dynamo. The *avalanche diode*, a voltage-conscious switch, allows reverse current to flow only when its breakdown voltage is reached. The *transistor*, a relay, allows current to flow in a collector-emitter circuit only when a small current is applied to the base-emitter circuit. An avalanche diode is used to provide the reference voltage in the electronic regulator, replacing the voltage coil and spring-loaded armature, whilst the transistors eliminate the vibrating contacts.

These devices are sensitive to abnormal currents in either direction or heat and are readily damaged by a wrong connection, incorrect polarity or the incorrect use of test equipment, battery chargers or electric welders.

Alternators

In the motor vehicle alternator a rotating field system operates inside stationary generating windings. The rotor has, typically, six iron claws projecting from each end, interlacing to form alternate north and south poles. These are magnetized by the field winding receiving current from two slip rings and brushes (Figure 15.7).

The stator has a circular laminated iron core carrying three separate output windings. Alternating current is induced in each of these windings by the revolving magnetic field. This three-phase alternating current is full-wave rectified by six silicon diodes mounted on the end plate in the airflow of the ventilating fan. Three further diodes provide direct current to the field winding controlled by a built-in microcircuit voltage regulator (Figure 15.8). No cut-out is needed because of diode action, and the alternator is self-regulating in current output owing to

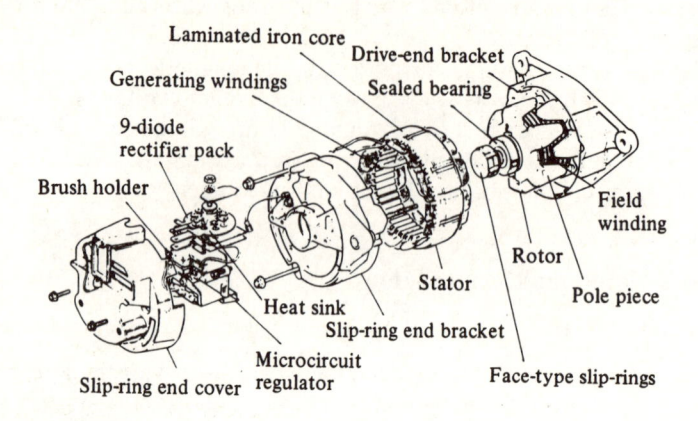

Figure 15.7 Alternator: exploded view

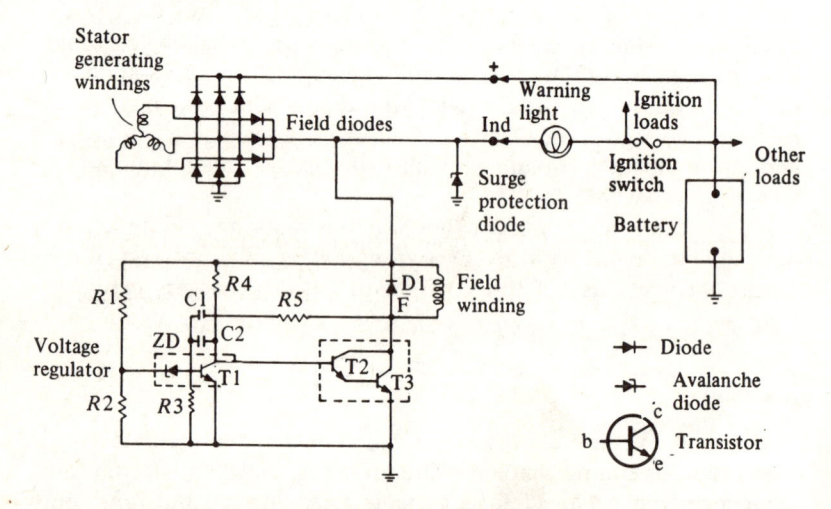

Figure 15.8 Alternator: wiring layout

the increasing reactance – inductive resistance – produced by the increasing frequency of the alternating current.

When the ignition switch is turned a small battery current flows through a warning lamp bulb to the alternator field winding and provides the initial magnetism until generation commences.

Although more costly, the alternator is inherently more robust and more reliable and can operate at about a third higher rev/min than the

The charging and starting systems

dynamo. There is an effective charge from idling speeds and the output to weight and output to space ratios are higher.

Microcircuit voltage regulator

The action of the regulator is as follows (Figure 15.8):

1. When the ignition is switched on a small current flows through the warning lamp and resistance R4 to the base-emitter circuits of transistors T2 and T3.
2. The collector-emitter circuit of T2, now conductive, allows current to flow from the alternator field winding to the base-collector circuit of T3.
3. The collector-emitter circuit of T3, now conductive, allows substantial battery current to flow through the field winding and the warning lamp illuminates.
4. As the alternator output increases, the three field diodes supply the field windings and the voltage rises.
5. When the output voltage reaches some 14 V, the proportional voltage applied to the avalanche diode ZD from R1-R2 is sufficient to make it conductive in the reverse direction. Current flows to the base-emitter circuit of T1.
6. The collector-emitter circuit of T1, now conductive, allows current from R4 to flow to earth. This switches off T2, T3 and thus the field current. The diode D1, by-passing surges induced by the collapsing field current, protects the transistors.
7. The alternator output voltage falling, the avalanche diode ZD is no longer conductive and the cycle repeats.

15.5 The starter motor

The action of an electric motor is the reverse of dynamo action since, when a current is passed through a conductor lying in a magnetic field, the interaction of this field with the one produced by the conductor tends to cause a movement of the conductor, in accordance with Fleming's left-hand rule.

Strip copper, or aluminium, field and armature windings are connected in series by copper-carbon brushes. The series type of motor, having its maximum torque at the slowest speed, is very suitable for starting purposes, and a current of 200–800 A is required depending on type.

The field windings may be four conventional coils, each pair in series between the supply and armature, while two earth brushes complete the

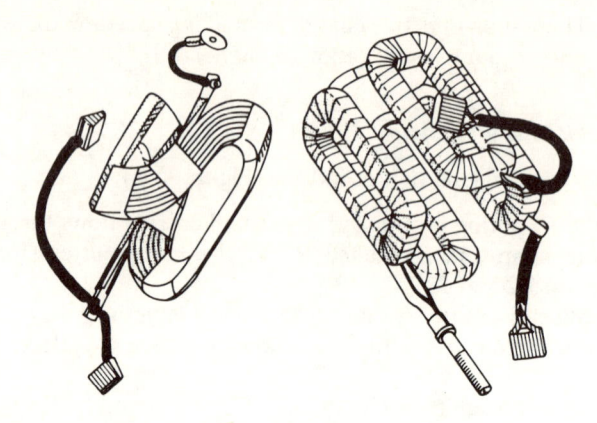

Figure 15.9 Conventional and wave-wound field coils

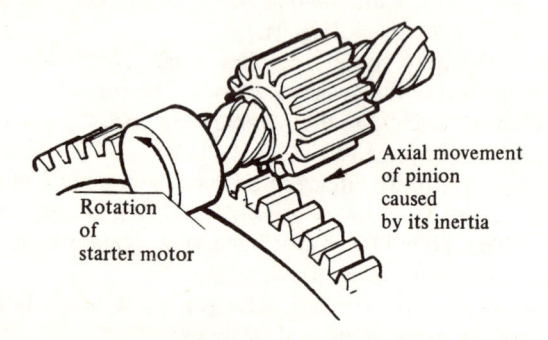

Figure 15.10 Inertia engagement

armature circuit. Alternatively two brushes may feed one end of a continuous wave-form field winding, the other end being earthed. Two feed brushes then complete the armature circuit (Figure 15.9).

The armature spindle may carry an inertia-engaged drive arrangement consisting of a pinion fitting on a screwed sleeve (Figure 15.10). When the motor is operated, the inertia of the pinion causes it to lag and move along the sleeve into mesh with the flywheel ring gear, a coil spring being employed to reduce the shock of engagement. When the engine fires, the ring gear rotates the pinion faster than the starter motor and drives it out of mesh.

The inertia-engaged starter motor is largely replaced by pre-engaged types (Figure 15.11). Usually a solenoid-operated pivoted lever slides the pinion into mesh with the ring gear before the electrical contacts

The charging and starting systems

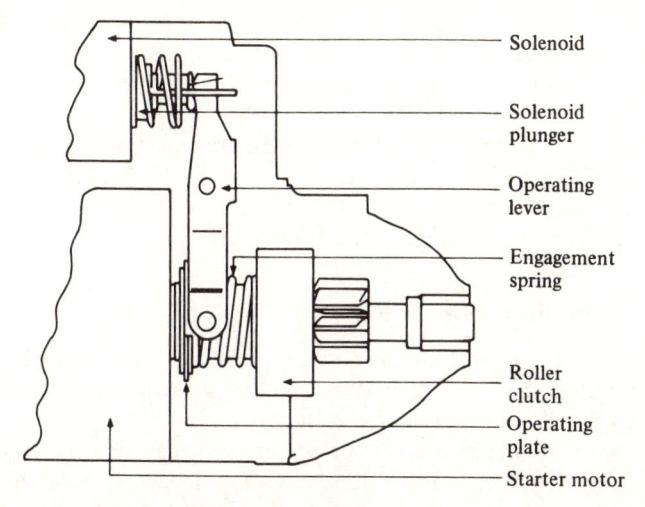

Figure 15.11   Pre-engaged starter motor

provide the starting torque. End-on contact between the teeth can be overcome by spring loading the pinion or using a partial field to slowly rotate the armature – low-power indexing – and only applying full power after complete engagement. A jamming-roller freewheel or a small multiplate clutch prevents the flywheel ring gear driving the starter armature.

Maintenance

Any accumulation of brush dust should be removed from the instrument since it may form a conducting track; if this becomes incandescent, it will burn the insulation. Slip rings or commutator can be cleaned with a petrol-moistened cloth. Only very limited skimming – in the lathe, using a sharp tool and fine feed – is possible with the moulded type of commutator, if ridged or worn. New brushes can be bedded to the curvature of the commutator using a strip of glasspaper between the two surfaces.

Where aluminium-strip starter field coils are employed, new brush leads can be soldered only to a copper stub of the old brush lead. Starter motors must not be tested without load; if run free they will attain a speed where the revolving conductors can be dislodged by centrifugal force. Should the inertia-type starter pinion jam during engagement, the armature spindle may be rotated from a squared extension at the commutator end.

Most electrical machines have either grease-packed bearings or oil-impregnated bushes. If lubrication is provided, care must be taken to avoid excess; pure oil is an insulator, but with carbon dust it forms a conductor.

Alternator or dynamo belt tension must prevent slip but not overload bearings. The longest run should have some 15–20 mm play under moderate finger pressure. The V-belt must not bottom in the pulley groove; this would prevent the wedging action.

15.6 Solenoid starter switch

As currents of several hundred amperes are required to operate the starter motor, the cables should be as short as possible and of adequate cross-sectional area to minimize voltage drop.

The switch contacts must be of adequate size and brought together under considerable pressure. Separate solenoid switches are therefore employed with inertia-engagement starter motors. On pre-engaged starters the solenoid is integral with the engagement system.

Chapter 16
The clutch

A clutch is a device for connecting or disconnecting two shafts. It may be a dog clutch, where the engagement is positive through projections on one member mating with corresponding indentations on the other, or a friction clutch, where the torque is transmitted through friction surfaces which allow a progressive engagement as they are brought together.

A friction clutch is necessary between the engine and gearbox to

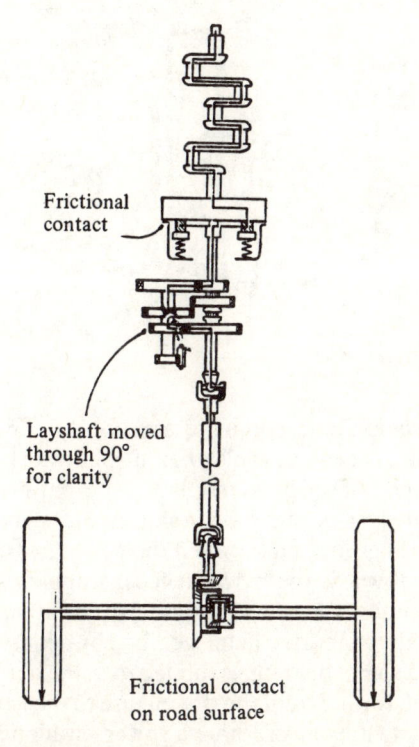

Figure 16.1   Transmission of power

disengage and permits a smooth and gradual re-engagment of the drive. When engaged, the clutch must transmit the maximum engine torque without slipping, and when disengaged for gear changing it must not exert a drag on the idle member (Figure 16.1).

The internal combustion engine must revolve at a reasonable speed to develop sufficient torque to move the vehicle from rest and, when starting, this engine torque must be smoothly and gradually conveyed to the gearbox without shock to the transmission.

16.1 The single-plate dry clutch

The driving members of the clutch consist of the flywheel and the pressure plate, both being of cast iron. The pressure plate is caused to revolve with the flywheel by three or four sets of tempered-steel straps arranged tangentially between the pressure plate and the cover bolted

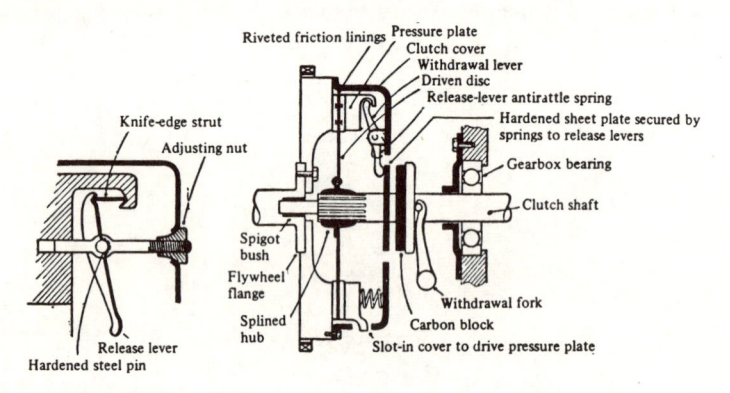

Figure 16.2 Clutch layout

on to the flywheel. These eliminate all rubbing contact between the pressure plate and the cover, whilst their flexibility in bending allows for the axial movement of the pressure plate. Alternatively, projections on the pressure plate engage and slide in slots in the cover (Figure 16.2).

A series of springs located between the cover and the pressure plate force the latter towards the flywheel face, trapping the friction-lined driven plate or clutch disc between these two surfaces.

The clutch shaft, supported in the gearbox by a journal bearing and in the flywheel by a spigot bearing, is splined to fit the hub of the clutch disc and transmits the torque from the driven disc to the gearbox.

The driven plate may incorporate a spring-cushioning hub to reduce shock loading during engagement and absorb torsional vibration. The

The clutch

hub springs can be arranged to have a progressive action by varying rates or end clearances, and some form of friction damping can be provided between the hub and the plate to absorb the torsional energy.

The riveted-on linings are of bonded asbestos and have a coefficient of friction on the cast-iron surfaces of 0.3–0.4. The spring-steel leaves carrying the linings may be arranged to slightly separate the two rings of material when free, so as to give a smooth action as they are pressed together during engagement. Slotting the plate eliminates heat distortion.

16.2 Withdrawal mechanism

Mechanical operation

The clutch is freed by withdrawing the pressure plate from the driven disc against the force of the clutch springs. This is accomplished by three or four release levers, pivoting on the clutch cover and engaging with the pressure plate at their outer ends, and the clutch withdrawal bearing on their inner ends. Knife-edge struts are used between the ends of the release levers and the pressure plate to reduce friction (Figure 16.3).

The clutch pedal may actuate the withdrawal bearing either by a pivoted fork lever or through a cross-shaft carrying a fork (Figure 16.4). The linkage from the clutch pedal has to accommodate engine movement on its flexible mountings caused by torque reaction as the clutch is

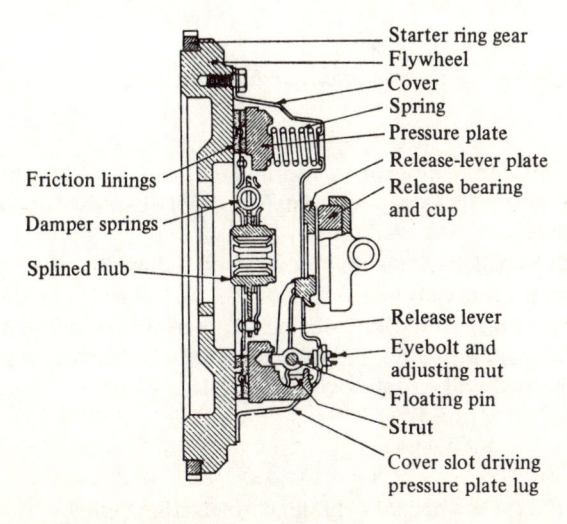

Figure 16.3 Multispring single-plate clutch

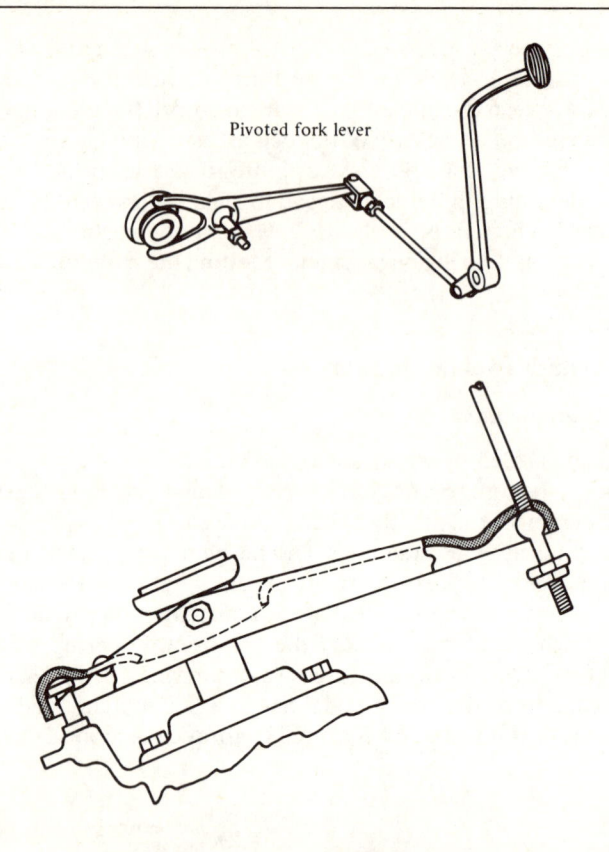

Figure 16.4   Mechanical operation: pivoted fork levers

engaged. This can complicate a rod linkage, and flexible-cable operation is frequently employed, with a nylon- or PTFE-lined outer casing to reduce friction.

The withdrawal bearing may consist of a carbon ring, in a steel carrier, which bears on to a hardened steel plate carried by the release levers. More usual is a grease-sealed deep-groove or angular-contact ball-race housed in a hub sliding along the sleeve extension of the gearbox clutch-shaft bearing cover (Figure 16.5).

Hydraulic operation

The pedal operates a master cylinder connected by pipeline and flexible hose to a slave cylinder actuating the clutch release fork. The system

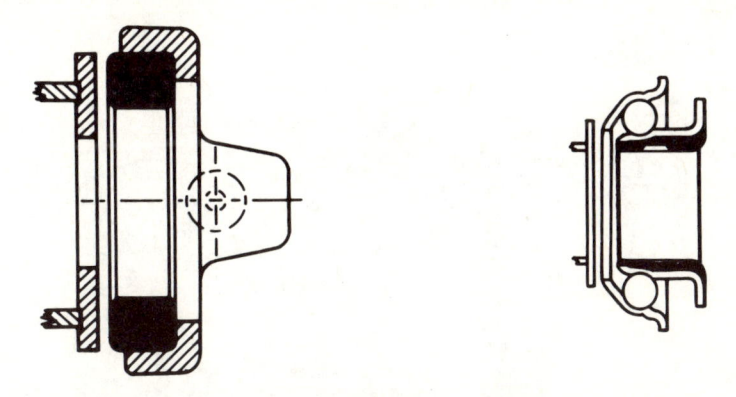

Figure 16.5 Clutch withdrawal bearings: graphite-impregnated ring; angular-contact ball bearing

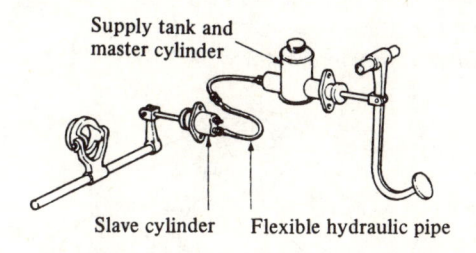

Figure 16.6 Hydraulic operation: cross shaft

reduces friction and eliminates any effect of torque reaction on the pedal movement (Figure 16.6).

16.3 Diaphragm-spring clutch

In the diaphragm-spring clutch the coil springs and the release levers are replaced by a Belleville-type spring-steel ring, slightly coned when free, flat when providing the clamping load, and coned in the reverse direction during withdrawal (Figure 16.7).

The diaphragm spring pivots on two fulcrum rings held by shouldered rivets to the clutch cover. Alternatively, the rings may be secured by integral tongues on the cover pressing, which are clenched over during manufacturing assembly. The outer rim of the diaphragm spring is retained in the pressure plate and its inward-pointing tapered fingers are acted upon by the release bearing (Figure 16.8).

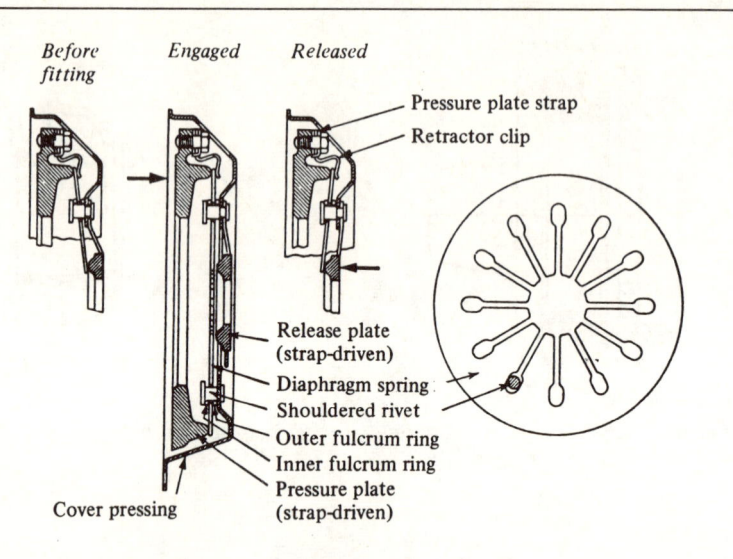

Figure 16.7   Diaphragm-spring clutch

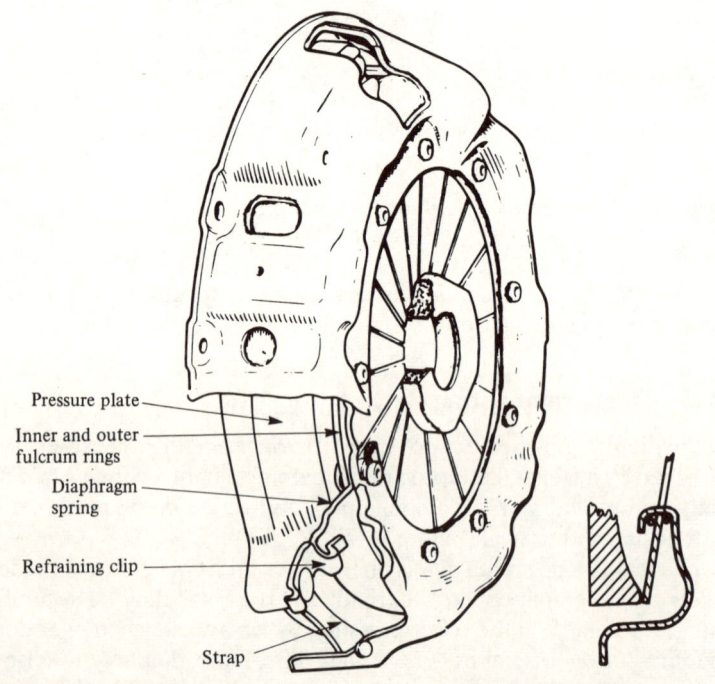

Figure 16.8   Details of diaphragm-spring clutch: strap drive; alternative fulcrum-ring retention

The clutch

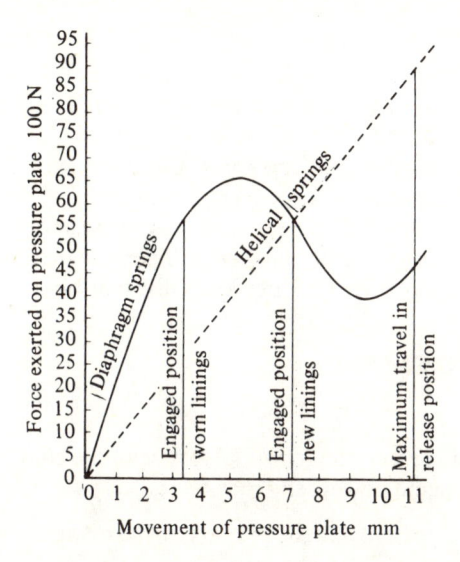

Figure 16.9 Load–deflection graph: multispring and diaphragm-spring clutch

The force produced by a diaphragm spring is not directly proportional to the deflection, i.e. it does not follow Hooke's law. Consequently clutch-pedal forces can be reduced whilst the clamping-spring pressure remains almost constant throughout the life of the linings (Figure 16.9). Other advantages of the diaphragm-spring clutch are a simplified construction, a more even circumferential clamping force, an independence of centrifugal force (unlike coil springs), and a more accurate balance.

16.4 Maintenance

As the friction linings wear down, the release levers move towards the withdrawal bearing and would eventually prevent the clutch from fully engaging. With mechanical operation a free movement of about 25 mm at the clutch pedal, or about 1.5 mm at the withdrawal bearing, is therefore provided to prevent clutch slip from this cause.

With hydraulic operation two systems are used, but in each case a slight free movement of about 4 mm must exist on the clutch pedal to ensure the complete return of the master-cylinder piston. The slave-cylinder piston rod may be adjustable to provide 1.5 mm clearance at the release bearing; alternatively a self-adjusting no-clearance system may be employed where a light rubbing contact exists between the release-lever plate and the bearing when the clutch is fully engaged.

16.5 Overhaul

The various parts of the clutch must be marked on removal to ensure their replacement in the same relative positions to maintain the original balance of the assembly.

Accurate resetting of the release levers is essential for satisfactory operation, and this requires a gauge plate or special setting fixture. Most diaphragm clutches can only be renewed as an assembly.

The gearbox must not be allowed to hang on the clutch during its removal or refitting otherwise the clutch disc may be buckled.

16.6 Clutch defects

Clutch slip

Clutch slip occurs when the frictional torque of the clutch is inadequate to convey the engine power, and may be caused by

(a) Insufficient free movement at clutch pedal or withdrawal bearing
(b) Worn linings
(c) Gearbox oil or grease on friction linings
(d) Scored faces of flywheel or pressure plate
(e) Weak or broken clutch springs.

Clutch drag

Clutch drag will produce difficulty in engaging or changing gear, and may be caused by

(a) Insufficient effective pedal travel – excessive free movement on pedal or slave-cylinder push rod
(b) Hydraulic system defective
(c) Withdrawal bearing defective
(d) Clutch disc buckled or cracked
(e) Release levers incorrectly adjusted or broken
(f) Oil or grease on friction surfaces
(g) Clutch disc seized on splines, or frozen (bedded) to friction faces after prolonged standing
(h) Clutch-shaft spigot bearing in flywheel seized.

Clutch judder

Sometimes vibration is produced as the clutch is engaged or, instead of a smooth progressive action, the vehicle jumps forward. The causes may be

The clutch

(a) Small quantity of gearbox oil on friction linings
(b) Defective release mechanism
(c) Engine mountings defective, or torque stay broken, allowing excessive torque reaction movement
(d) Loose or worn linings, loose rivets
(e) Excessive backlash in worn universal joints or transmission.

Apart from mechanical defects, incorrect driving methods may cause clutch defects, as, for example, by resting the foot on the clutch pedal.

Example 19

An engine develops 146 Nm torque which is transmitted through a single-plate clutch of mean diameter 250 mm. If the coefficient of friction between the surfaces is 0.3, calculate the minimum total clutch-spring force required to transmit this torque.

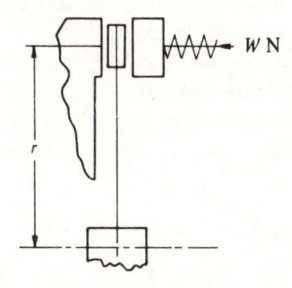

Figure 16.10

Let W N be the minimum total spring force (Figure 16.10). This is the normal force between the surfaces. The frictional force for each surface of the clutch plate $= \mu W$ N, and the total frictional force for the two sides of a single-plate clutch $= 2\mu W$ N. Hence

$$\text{torque transmitted} = 2\mu W r \quad \text{Nm}$$

where r = mean radius (m). Then

$$W = \frac{\text{torque transmitted}}{2\mu r} \quad \text{N}$$

$$= \frac{146}{2 \times 0.3 \times 0.125}$$

$$= 1946.7 \text{ N}$$

The minimum total clutch-spring force required to transmit the torque without slip will be 1946.7 N.

215

Chapter 17
The gearbox

17.1 Tractive resistance

When a vehicle is travelling at constant speed, its resistance to motion, termed the tractive resistance (Figure 17.1), consists of:

Rolling resistance This depends mainly upon the nature of the ground, the tyres used, the weight of the vehicle and, to a lesser extent, the speed (the last variation is usually ignored).

Wind resistance Wind resistance depends upon the size and shape of the vehicle – its degree of streamlining – and increases approximately as the square of the speed through the air.

Gradient resistance This is determined by the steepness of the hill and the weight of the vehicle, which must, in effect, be lifted from the bottom to the top.

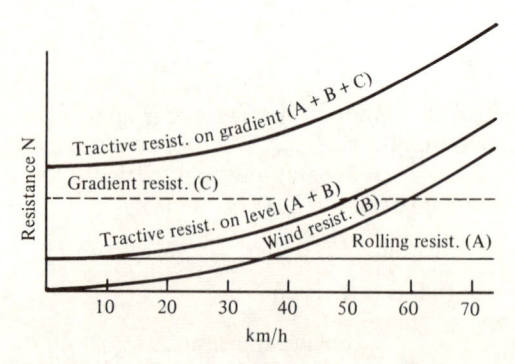

Figure 17.1 Tractive resistance curves

17.2 Tractive effort

For a uniform speed the tractive resistance must be balanced by the tractive effort or *driving force* produced at the point of contact of the tyres on the road by the driving torque.

The gearbox

For acceleration, the driving force must be greater than the total tractive resistance in order to overcome the inertia of the vehicle.

17.3 Gear ratio

The gear ratio, or velocity ratio, between a pair of gear wheels is in inverse ratio to the number of teeth on each. Thus, for Figure 17.2,

$$\frac{\text{rev/s of A}}{\text{rev/s of B}} = \frac{\text{number of teeth on B}}{\text{number of teeth on A}}$$

This may be extended to deal with pulley and belt drive (presuming no slipping):

$$\frac{\text{rev/s of A}}{\text{rev/s of B}} = \frac{\text{circumference or diameter or radius of B}}{\text{circumference or diameter or radius of A}}$$

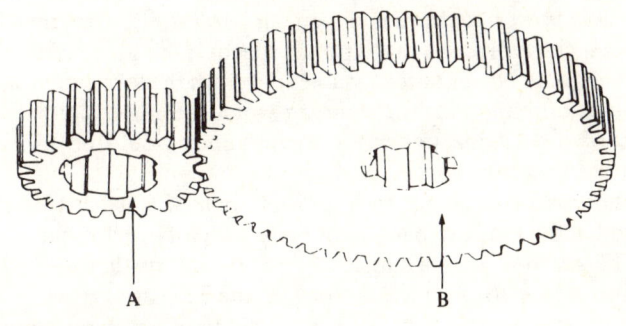

Figure 17.2 Gear ratio

17.4 Power, speed and torque

The power transmitted by a shaft is directly proportional to the speed of revolution and the torque acting on it:

power (W) = $2\pi \times$ torque (Nm) $\times$ rotational speed (rev/s)

For a given power, therefore, the torque is inversely proportional to the speed of revolution, and if the rev/s is reduced the torque will be increased in the same ratio.

17.5 Front- and rear-wheel drive

The maximum driving torque that can be employed is limited by the force of friction between the tyres and the road:

$$F = \mu W$$

where F = force of friction between the tyres and road, μ = coefficient of friction (very variable with conditions, say 0.8 as a good average), and W = normal force on tyres.

A tractive effort greater than the force of friction will produce wheel spin.

During acceleration there is a weight transfer to the rear wheels, and the normal force between the tyres and the ground from the static weight distribution is increased at the rear and reduced at the front. The degree of accompanying body *squat* is determined by the suspension layout.

17.6 Gearbox and final-drive ratios

As the internal combustion engine develops its power at comparatively high rev/min, a gear reduction, and a torque increase, is necessary to enable this power to be developed and transmitted to the driving wheels, which revolve at a much slower speed. If this gear ratio is chosen to give adequate power at reasonable speeds on the level, then starting, slow-speed running and hill climbing would be impossible since the rev/min of the engine under these conditions would be too low to produce the power required.

For these reasons a gear reduction of about 4:1 or 5:1 (for cars) is employed in the final drive to provide a satisfactory performance on the level. The gearbox gives a choice of three or four further reductions to allow the engine to produce the power and increased driving torque necessary for the varying conditions of starting, accelerating and hill climbing. The gearbox also provides a neutral position, allowing the engine to run with the clutch engaged and the vehicle stationary, and a means of reversing the drive to the road wheels.

Bottom and top gear are those with respectively the largest and smallest gear ratios. Engaging a numerically higher ratio in the gearbox is called 'changing down to a lower gear'. Similar possible confusion applies to the final drive ratios: a high-ratio axle is one providing a low numerical gear ratio.

17.7 Curves of driving torque

Since, for a given power, the torque and rev/min are inversely proportional, the curves show the increase in driving torque for the lower gear ratios with a corresponding reduction in the range of road speed.

By superimposing a curve of tractive resistance on a gradient –

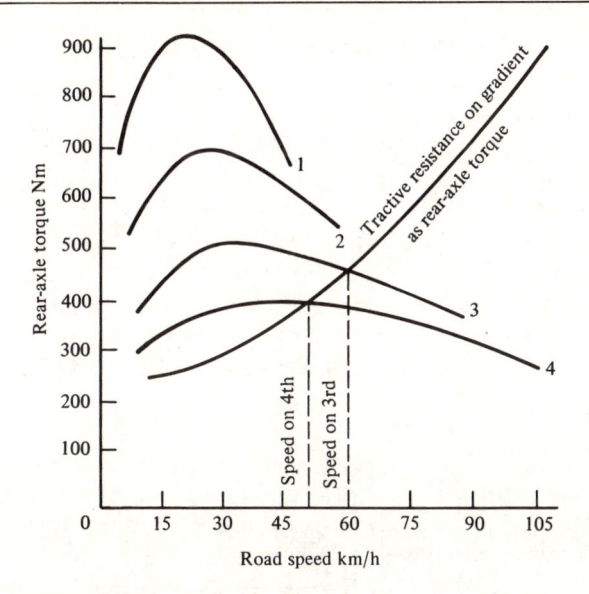

Figure 17.3 Driving-axle torque curves

expressed in terms of driving torque – it can be seen that the vehicle will make a faster climb in the lower gear (Figure 17.3).

By plotting the power available at the rear wheels and the tractive power required on the level (where tractive power on the level = tractive resistance × speed in m/s) the relationship of the engine power to the vehicle speed can be shown (Figure 17.4).

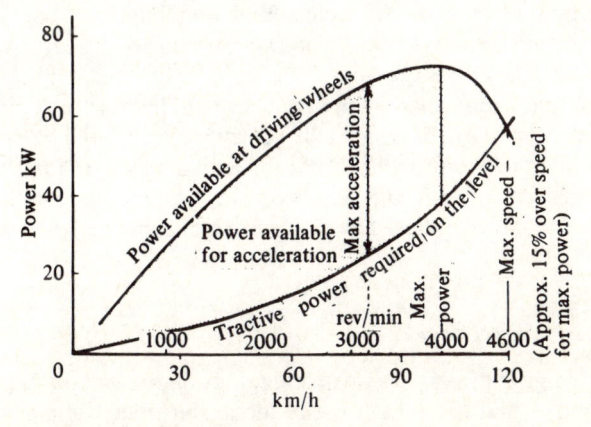

Figure 17.4 Power requirements

17.8 The sliding-mesh gearbox

The gearbox casing consists of an aluminium-alloy or malleable-iron casting extended at the front to form the clutch housing, or alternatively having a separate bolted-on housing. The rear of the casing carries the engine mountings, where these are not fixed to the clutch housing or the rear of the engine.

The engine torque is conveyed from the driven disc of the clutch by the clutch (primary or first-motion) shaft. The clutch shaft revolves in a bearing in the gearbox casing and has an integral pinion which is in permanent engagement with a corresponding pinion on the layshaft below it, the two being termed the constant-mesh pinions (Figure 17.5).

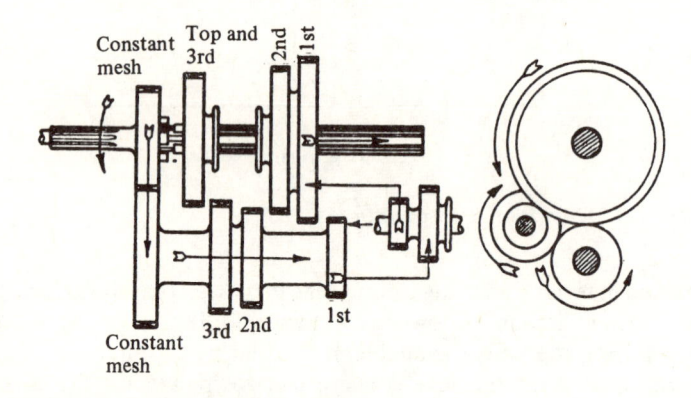

Figure 17.5 Four-speed sliding-mesh gearbox

The layshaft consists of a cluster of four pinions – constant mesh, third, second and first gear – rotating on a fixed layshaft spindle or in bearings in the gearbox casing.

The splined mainshaft is carried in a spigot bearing in the clutch shaft at the front, and in a bearing in the gearbox casing at the rear end, where it is coupled to the propeller shaft through a universal joint. Sliding on the mainshaft is the third-gear pinion and a combined first- and second-gear pinion; both of these have integral collars into which fit the selector forks.

Top gear or direct drive

The third-gear pinion has projecting dogs on the side facing the clutch shaft and, when the pinion is slid along the splined mainshaft by the selector fork, these dogs mate with corresponding projections on the clutch shaft to give a positive direct drive between the two shafts for top gear.

Third gear

For third gear the mainshaft pinion is slid into engagement with the third-gear pinion on the layshaft. The drive is indirect and passes from the clutch shaft through the constant-mesh (CM) gears to the layshaft and back to the mainshaft through the third-gear pinions. The gear ratio is the product of the gear ratios of the two pairs of pinions:

gearbox ratio, third gear

$$= \frac{\text{rev/s of clutch shaft}}{\text{rev/s of mainshaft}}$$

$$= \frac{\text{rev/s of clutch shaft}}{\text{rev/s of layshaft}} \times \frac{\text{rev/s of layshaft}}{\text{rev/s of mainshaft}}$$

$$= \frac{\text{no. teeth on layshaft CM pinion}}{\text{no. teeth on clutch-shaft CM pinion}}$$

$$\times \frac{\text{no. teeth on mainshaft third gear}}{\text{no. teeth on layshaft third gear}}$$

This is more easily remembered by taking the pinions in pairs in their order from the mainshaft to the clutch shaft:

gearbox ratio

$$= \frac{\text{teeth on mainshaft pinion}}{\text{teeth on corresponding layshaft pinion}}$$

$$\times \frac{\text{teeth on layshaft CM pinion}}{\text{teeth on clutch-shaft CM pinion}}$$

Second gear

In this case the integral second-first pinion on the mainshaft is engaged with the corresponding second-gear pinion on the layshaft, and owing to their respective sizes a greater gear reduction is obtained than with the third-gear pinions. The ratio is calculated in a similar manner, using the number of teeth on the second-gear pinions.

First gear

The integral second-first pinion is slid into engagement with the first-gear pinion on the layshaft and a further gear reduction is obtained.

Reverse gear

The reverse pinion has a separate selector fork which moves it into mesh with the first-gear pinions on the layshaft and mainshaft when these are in the neutral position. A spring-loaded plunger or similar device is frequently incorporated in the reverse selector position to avoid accidental engagement. If the reverse ratio is to be lower than the first-gear ratio, then a combined gear must be used having the larger pinion meshing with the layshaft and the smaller pinion engaging with the mainshaft gear. The reverse ratio can then be calculated as:

$$\text{reverse ratio} = \frac{\text{no. teeth on mainshaft first gear}}{\text{no. teeth on reverse 'mainshaft' pinion}} \times \frac{\text{no. teeth on reverse 'layshaft' pinion}}{\text{no. teeth on layshaft first gear}} \times \frac{\text{no. teeth on layshaft CM gear}}{\text{no. teeth on clutch-shaft CM gear}}$$

17.9 Selector mechanism

In the four-speed gearbox, three selector forks are required to slide the mainshaft pinions and the reverse pinion. Each fork is carried on a selector rod housed in the gearbox casing or cover, and has a slot in its upper surface into which fits the lower end of the gear lever. The gear lever is ball-mounted, and transverse movement selects one or other of the slots in the selector forks. A longitudinal movement of the lever then slides the chosen fork and its pinion to engage the selected gear (Figure 17.6).

When the gearbox is remote, various methods of transmitting the selection and engagement movement from the gear lever can be used. Often a rod with twisting and sliding action is sufficient, but with transversely mounted gearboxes it may be necessary to use bell crank relay levers and links (Figure 17.7).

The selector forks are retained in their desired positions by spring-loaded balls or plungers fitting into V-shaped notches in the selector

The gearbox

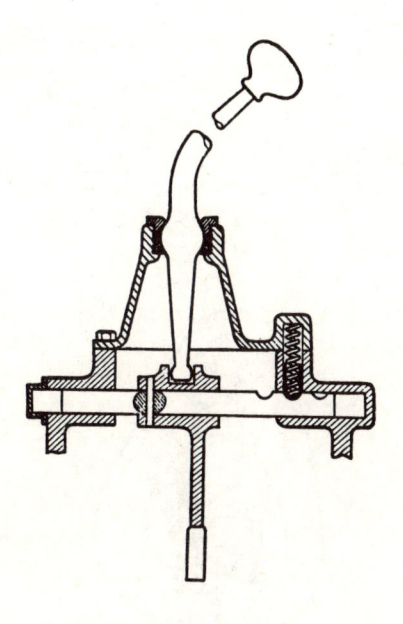

Figure 17.6   Gear-lever selector mechanism

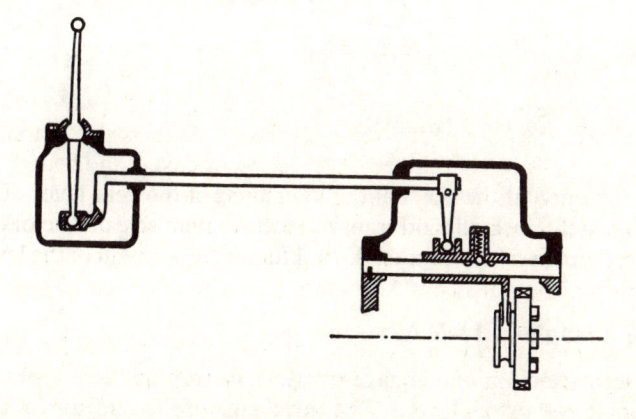

Figure 17.7   Remote-control gear change

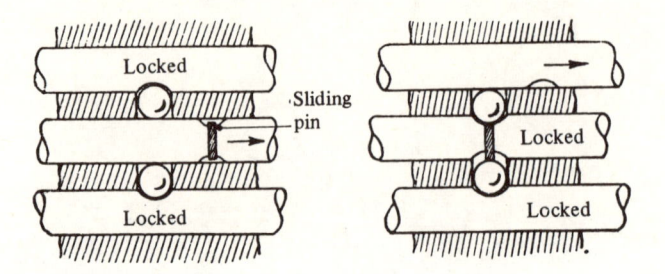

Figure 17.8 Ball interlock

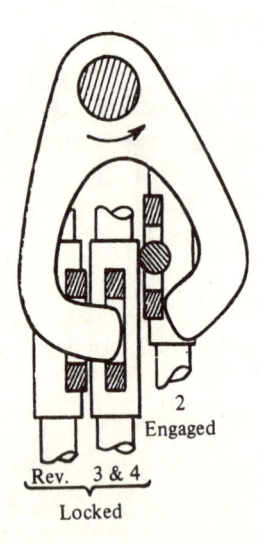

Figure 17.9 Caliper interlock

rods. To prevent the possibility of engaging two sets of gears at once – which would lock the drive – a selector interlock of the caliper or plunger type is fitted (Figures 17.8, 17.9).

17.10 Single-rail selection

The gear selection and engagement mechanism can be simplified into the single rod or rail layout. The three jaws of the selector forks are in line in the neutral position and are interlocked by a caliper swinging through them, as in the previous system. The innovation is that the two mainshaft selector forks slide along the single rail and the reverse-fork lever is pivoted on to the side of the gearbox (Figure 17.10).

The gearbox

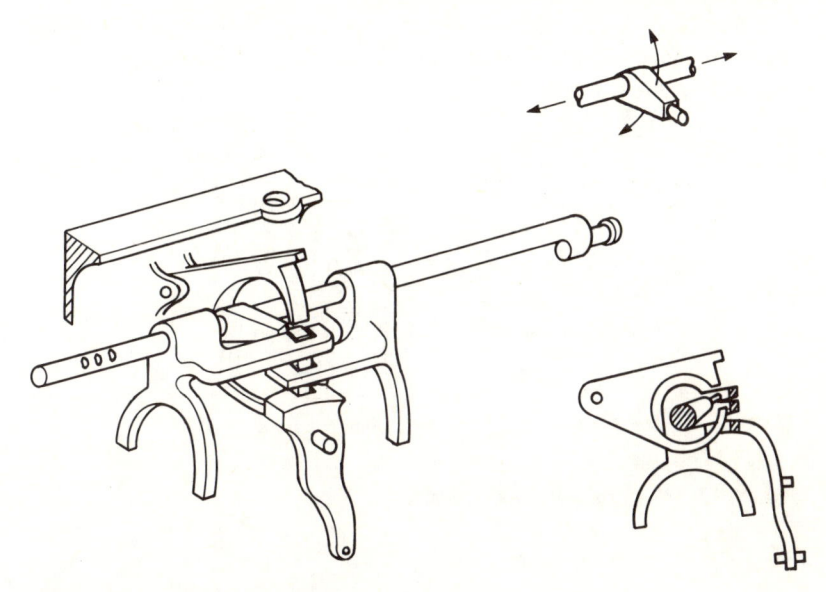

Figure 17.10 Single-rail selector

Fixed on the rail is an engagement striker which can be twisted into line with any one of the three selector jaws, and as this is done the caliper swings with the striker and locks the two unselected forks. Axial movement of the rail then engages the selected gear.

To rotate and slide the rail, its rearmost end is cranked and engages with the forked lower end of the ball-mounted gear lever. At the forward end, a detent spring and ball engages with one of the three grooves in the rail to hold the selected fork in position.

17.11 The constant-mesh gearbox

In this design, the mainshaft pinions revolve freely on bushes or needle-roller bearings and are all in constant engagement with the corresponding layshaft pinions (Figure 17.11).

Sliding on the mainshaft splines are dog clutches, moved by the selector forks, which engage with corresponding dogs on the mainshaft pinions and so lock the selected pinion to the mainshaft to give the required gear ratio. Top-gear engagement is by dog clutch as before.

With this arrangement the quieter-running helical gears can be employed, and during gear changing the noise and wear are reduced by the simultaneous engagement of all the dogs instead of only a pair of gear teeth as on the sliding-mesh box.

225

With single helical pinions (double helical is economically impractical) the driving loads on the teeth cause an axial thrust which must be resisted by thrust washers, or shoulders, on the mainshaft.

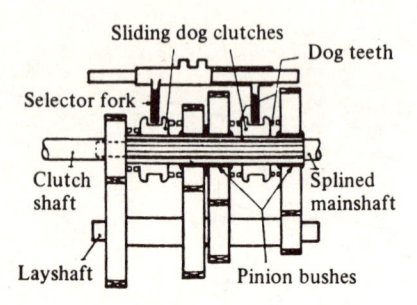

Figure 17.11   Constant-mesh gearbox

17.12   Gear changing

The operation of gear changing is similar with both sliding-mesh and constant-mesh gearboxes. The action with the latter type is described here.

When changing down, if the dog clutches are to engage without grating, the speed of the mainshaft pinion and the sliding dog must be equal. The sliding dog is splined to the mainshaft and its speed is determined by the speed of the car; therefore for the lower-gear ratio the speed of the clutch shaft, layshaft and mainshaft pinion must be increased, and this is accomplished by *double declutching*. In this operation the clutch is disengaged, the gear lever moved to neutral and the clutch re-engaged. The engine and engine-driven gearbox members are speeded up by a touch on the accelerator pedal. The clutch is disengaged, the gear lever moved to the lower gear and the clutch re-engaged.

When changing up, an opposite effect is required – a pause with the gear lever in neutral to allow the engine-driven members to slow down sufficiently for a noiseless engagement. To speed upward changes on a heavy vehicle, where the engine-driven members have considerable momentum, a brake may be applied to the clutch shaft as the clutch is fully withdrawn.

The success and quietness of these operations, particularly with heavy vehicles, depends upon the skill of the driver.

The gearbox

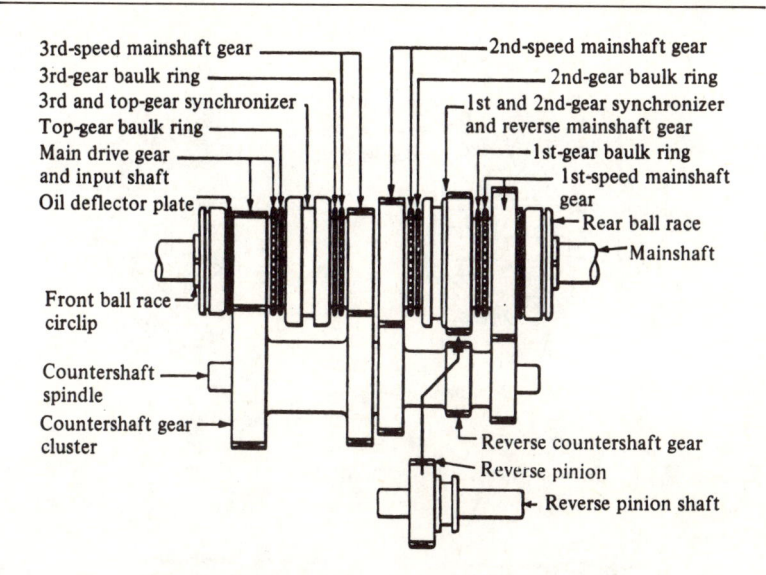

Figure 17.12 All-synchromesh four-speed assembly

17.13 The synchromesh unit

The synchromesh arrangement is a modification of the constant-mesh gearbox where the sliding-dog clutches and the corresponding gear pinions are provided with friction cones (Figure 17.12). The initial movement of the gear lever when the friction cones are pressed together provides the synchronizing action. For example, in changing down, the force of friction between the cones increases the speed of the mainshaft pinion, together with the layshaft and clutch shaft, until it is equal to the mainshaft speed. A continued movement of the gear lever then engages the dog clutch. By this means, noiseless gear changing is simplified and double declutching is unnecessary.

Various designs have been used where the force from the selector fork is transmitted to the friction surfaces through springs; sufficient compression of these allows an overriding action and positive engagement of the dog clutch (Figure 17.13).

Current units employ a baulking device to prevent a too rapid movement of the gear lever beating the cones – i.e. clashing the dog teeth together before synchronization is effected. The usual arrangement is a floating ring which locks the sliding-dog clutch member until the speeds are equalized (Figures 17.14, 17.15).

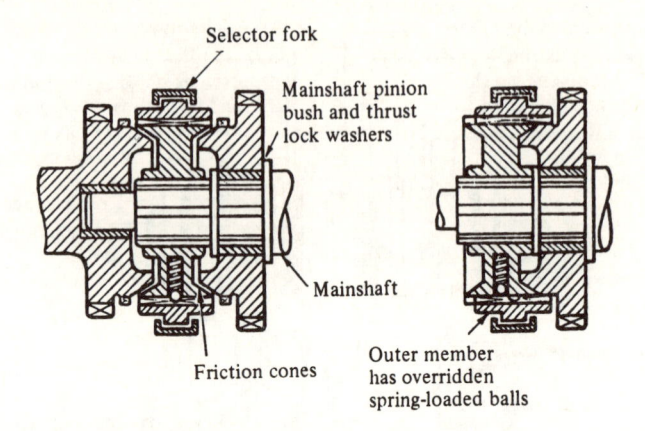

Figure 17.13 Spring-load synchromesh

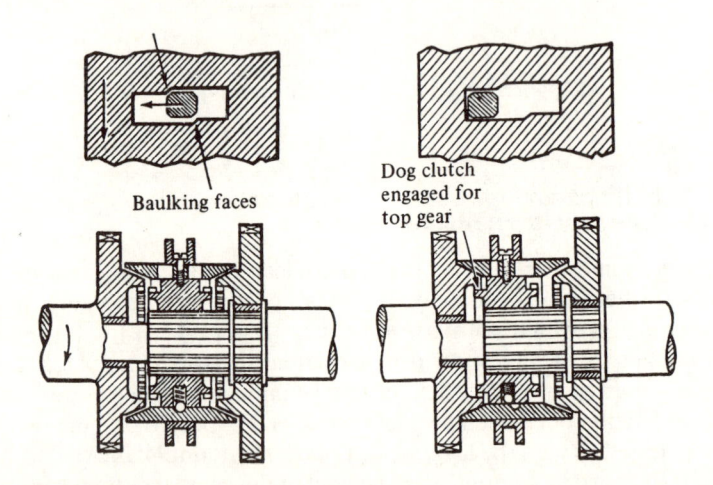

Figure 17.14 Baulk-action synchromesh

17.14 Additional gear ratios

Commercial vehicles having a relatively low power/weight ratio, and operating under unladen to fully loaded conditions, require additional gears for efficient operation.

One arrangement is to provide two pairs of alternative-ratio constant-mesh gears between the clutch shaft and layshaft. This doubles the number of indirect gear ratios available.

The gearbox

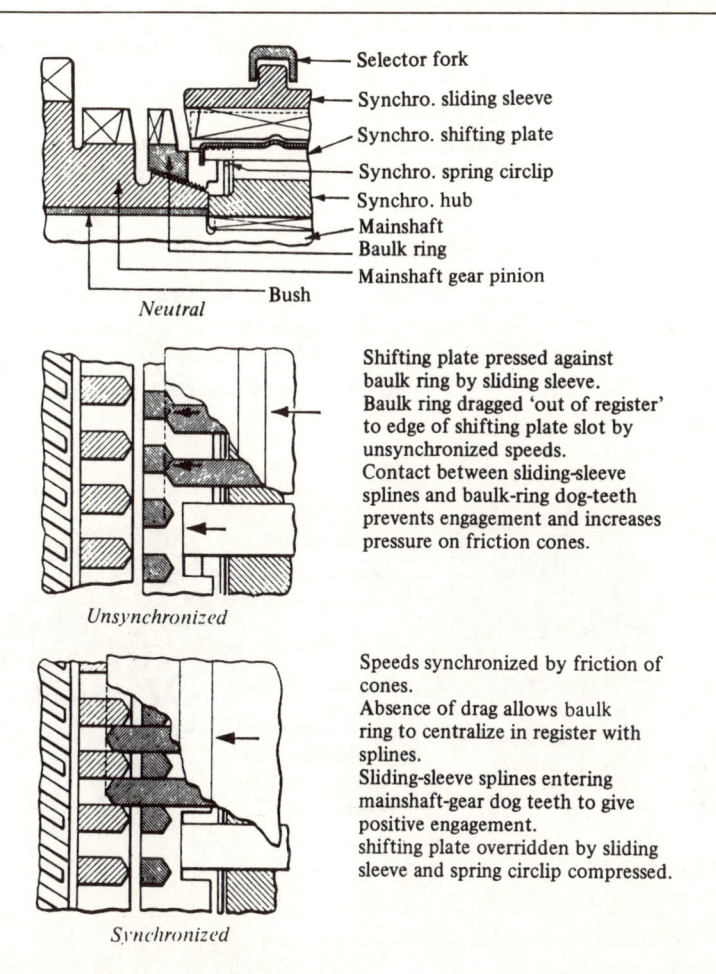

Neutral — Selector fork, Synchro. sliding sleeve, Synchro. shifting plate, Synchro. spring circlip, Synchro. hub, Mainshaft, Baulk ring, Mainshaft gear pinion, Bush

Unsynchronized — Shifting plate pressed against baulk ring by sliding sleeve. Baulk ring dragged 'out of register' to edge of shifting plate slot by unsynchronized speeds. Contact between sliding-sleeve splines and baulk-ring dog-teeth prevents engagement and increases pressure on friction cones.

Synchronized — Speeds synchronized by friction of cones. Absence of drag allows baulk ring to centralize in register with splines. Sliding-sleeve splines entering mainshaft-gear dog teeth to give positive engagement. shifting plate overridden by sliding sleeve and spring circlip compressed.

Figure 17.15   Baulk-ring synchromesh

Another system is to use an auxiliary gearbox behind the main gearbox with a choice of direct drive or a reduction to split the ratios in the main gearbox. This enables all the available gears to be used in sequence. The auxiliary gearbox may be a layshaft type with constant-mesh gears, or epicyclic, and the gear change may be power-operated electrically or by compressed air.

Sometimes, and particularly for cars where economy with a lowered cruising engine speed is desired, the epicyclic unit may provide an overdrive of approximately 0.75:1. More recent practice is to incorporate the overdrive within the gearbox so that fourth speed is direct drive and fifth speed an indirect ratio of some 0.75:1 to 0.85:1. A typical arrangement is an extra pinion on the layshaft in constant mesh with a mainshaft pinion turning on needle-roller bearings. This is engaged by a synchromesh unit splined to the mainshaft and operated from the reverse selector rod.

17.15 Epicyclic overdrive

The gearbox mainshaft drives the planet carrier of an epicyclic gear train (Figure 17.16). Three planet wheels on the carrier engage with the annulus which drives the tailshaft and the sliding sun pinion. In direct drive, the sun pinion is locked to the planet carrier and the epicyclic train revolves as a unit.

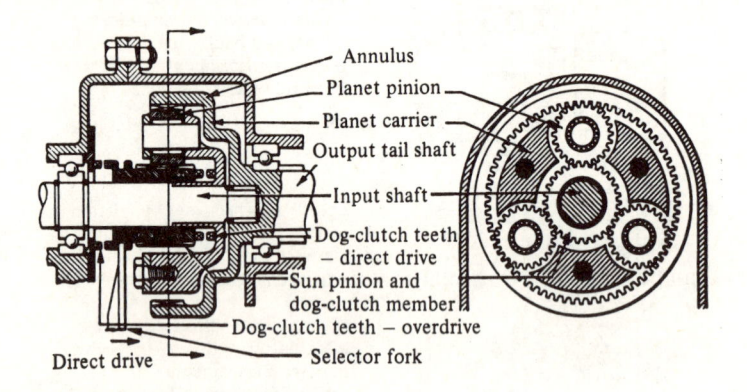

Figure 17.16 Epicyclic overdrive unit

For overdrive, the sun pinion is locked to the casing and the planet wheels revolving around the fixed sun drive the annulus at an increased speed.

In a typical epicyclic overdrive unit, hydraulically operated friction-cone clutches are used for gear engagement instead of dog teeth. A jamming-roller freewheel maintains the drive during the gear change, which is controlled electrically.

17.16 The freewheel unit

A freewheel unit can be fitted to the mainshaft behind the gearbox to prevent drive being transmitted from the propeller shaft to the mainshaft. The mechanism is usually of the jamming-roller type with a dog clutch to lock it out of action (Figure 17.17).

The gearbox

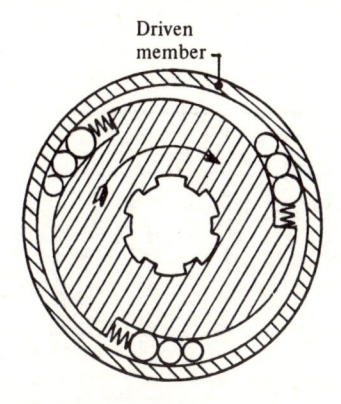

Figure 17.17   Jamming-roller freewheel

With the freewheel operating and the engine clutch disengaged, the gearbox is isolated from the engine and the drive and gear changing is greatly facilitated. The coasting effect increases petrol economy but, as the engine can exert no retarding force, driving methods must be modified and greater use of the brakes is probable. The dog clutch is automatically engaged by a connection from the reverse-gear selector and can be locked by a manual control for traffic conditions.

17.17   The all-indirect gearbox

The layshaft two-stage gearbox is used in both longitudinal- and transverse-engined front-wheel-drive cars. However, many of the former employ a single-stage, all-indirect gearbox (Figure 17.18).

The gearbox input shaft carries a number of integral or fixed pinions and is driven by a splined coupling from the clutch shaft. The driven shaft is integral with the final-drive pinion, which may be spiral bevel or hypoid type. This output shaft carries the meshing gears on needle-roller bearings, and the necessary synchromesh hubs. There is no direct drive and consequently no particular advantage in a 1:1 gearbox ratio.

17.18   Two-speed transfer gearbox

A range of vehicles uses optional four-wheel drive – with additional 'emergency' low ratios – to provide a cross-country facility. This is usually accomplished by a two-speed transfer gearbox, with layshaft and two pairs of constant-mesh helical gears, attached to the end of the main gearbox or driven via a short coupling shaft from the gearbox mainshaft

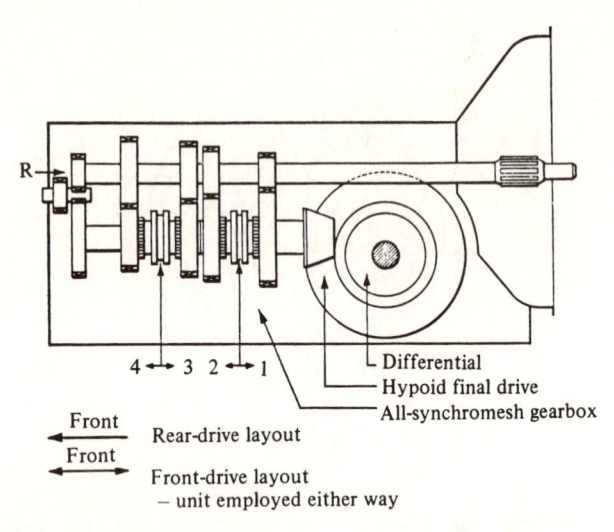

Figure 17.18 All-indirect gearbox

(Figure 17.19). Dog clutches engage the extra axle drive and select the high or low range; the latter is usually blocked except during 4WD to safeguard a single-axle transmission from excessive torque.

17.19 Lubrication

Gearbox lubrication is normally by oil transported by the gear teeth and aided where necessary by catch troughs and guide tubes. Bushes and bearings sometimes can be fed from radial drillings in gear-teeth roots. In a few cases force-feed lubrication is used, often by a crescent-type internal-gear pump feeding axial drillings in the shafts.

Oil retention for the clutch shaft and mainshaft may include oil flingers, scroll return threads and usually lip-type seals. Venting is necessary to avoid any pressure build-up.

Where the gearbox shares lubrication with the final drive, the choice of lubricant may be dictated by the load conditions for the latter gearing.

17.20 Gearbox defects

Noise in operation

(a) Shortage of lubricant
(b) Worn or pitted bearings

The gearbox

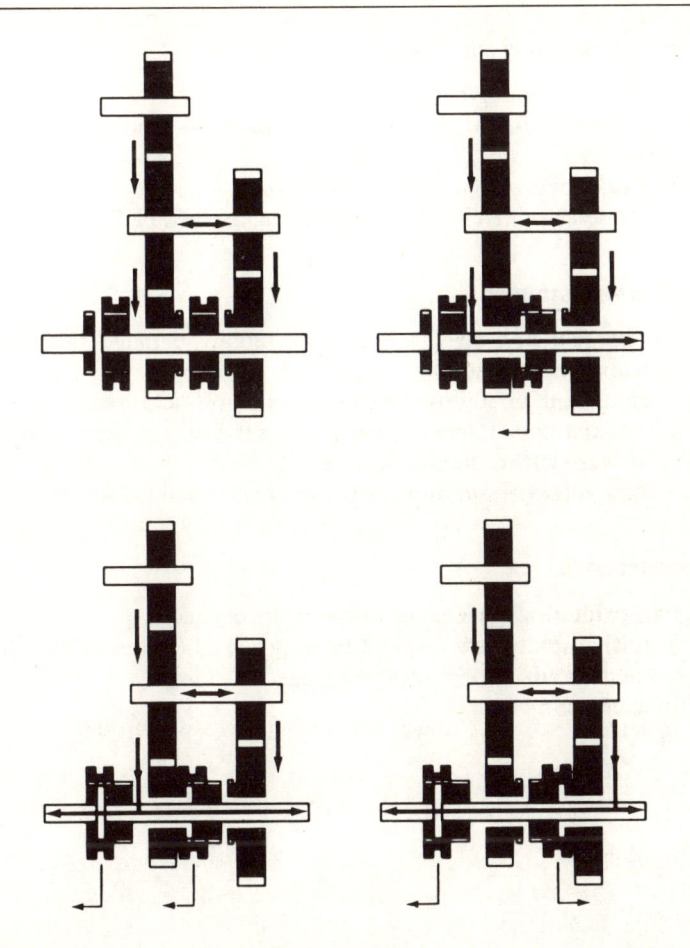

Figure 17.19 Two-speed transfer gearbox: no drive; high 2WD; high 4WD; low 4WD

(c) Incorrect meshing of gear teeth – worn bearings or bushes.
(d) Chipped or worn gear teeth.

Difficult engagement: constant-mesh gearbox

(a) Unskilful driver
(b) Engine clutch dragging
(c) Burred splines or teeth on gears or dogs
(d) Defective selector or selector-lock mechanism.

233

Gears jumping out of engagement

(a) Worn teeth or dog-clutch members
(b) Excessive end-float of shafts or pinions – worn bearings or locating washers
(c) Selector forks bent or incorrectly adjusted
(d) Worn selector mechanism, weak springs, seized balls or plungers.

Faulty synchronization

(a) Engine clutch dragging (this will rapidly damage the friction synchromesh surfaces)
(b) Incorrect oil, or additives affecting action of friction surfaces
(c) Worn friction surfaces; oil clearing threads worn; weak or broken springs in synchromesh unit
(d) Too rapid gear-lever movement, where no baulking device.

Jammed gearbox

(a) Faulty interlock mechanism – two gears engaged
(b) Mainshaft/clutch-shaft spigot bearing seized (shafts will turn only when direct drive or neutral is engaged)
(c) Broken components.

Chapter 18
Automatic transmission

18.1 Fluid flywheel or fluid coupling

The fluid flywheel comprises two rotors, each with radial vanes, in a casing filled with light engine oil. The driving member is attached to the crankshaft, and as this rotates the oil contained between the vanes is thrown outwards by centrifugal force and enters the opposing vanes of the driven member, which is connected to the gearbox. The oil passes to the inner portion of the driven rotor and returns to the driving members so that a circulation of oil is produced within the two rotors (Figure 18.1).

The driving member imparts energy of motion to the oil and some of this energy is passed to the driven rotor, so imposing an increasing drag on it as the engine speed rises. The slip, or difference in speed of the two rotors relative to engine speed, depends upon the engine speed and the load on the driven member. In general, as the engine speed rises from about 500 rev/min to 1000 rev/min the slip decreases from 100 per cent to some 10 per cent. This effect is used to eliminate the action of

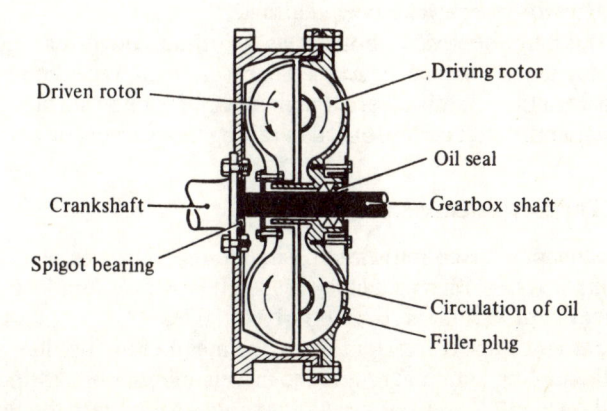

Figure 18.1 Fluid coupling

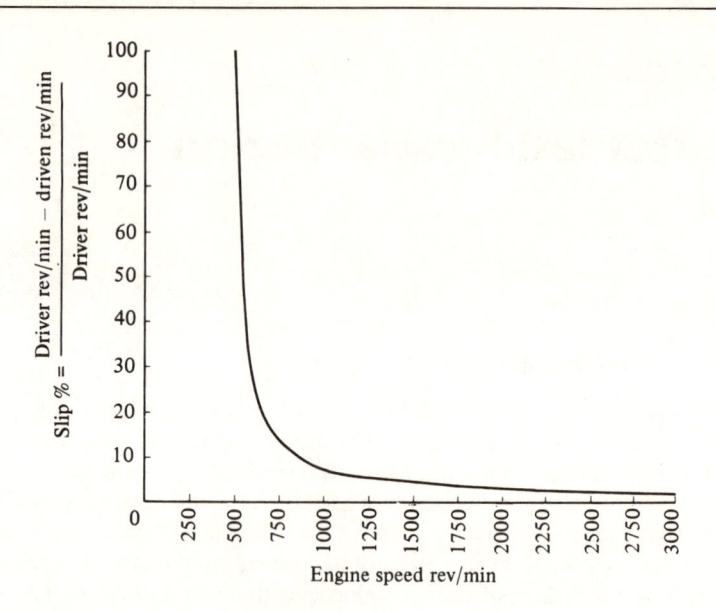

Figure 18.2 Fluid-coupling: slip curve

engagement of a friction clutch, and in addition the fluid flywheel serves to damp out torque fluctuations (Figure 18.2).

With a further increase in engine speed the slip falls to 1 per cent or 2 per cent, but there is inevitably some loss of energy and a resulting slightly lowered transmission efficiency. Even at the lowest speeds some drag is imposed on the driven rotor, making a simple combination of fluid flywheel and orthodox gearbox unsuitable, so that it is frequently employed with an epicyclic-type gearbox.

The fluid flywheel does not act as a torque converter, increasing engine torque as with a gearbox. The action may be compared to slipping a friction clutch except that no wear of friction surfaces results, and the heat produced is absorbed in raising the temperature of the oil.

18.2 The torque converter

The torque converter performs a similar function to the gearbox – increasing torque with a reduction in speed. However, unlike a conventional gearbox the ratio is continuously variable.

In the single-stage converter the driving member or impeller acts as in the fluid coupling and fluid enters the driven member or turbine. There is now, however, a stator or reaction member to redirect the flow from the turbine into the direction of engine rotation before it re-enters the

Automatic transmission

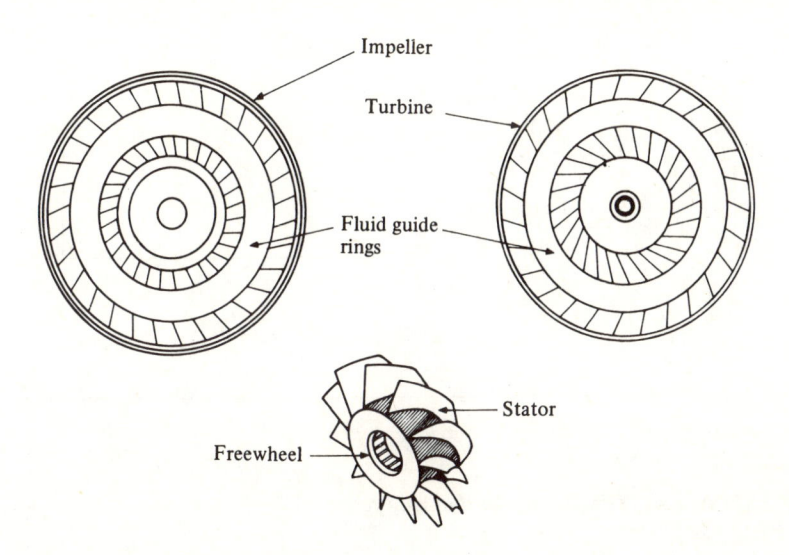

Figure 18.3 Torque converter

eye of the impeller. The energy remaining in this returning fluid, assisting the impeller flow, provides the torque multiplication (Figure 18.3).

Torque multiplication is at a maximum with the turbine stalled and falls to zero when both members revolve at the same speed. At both these extremes the efficiency is zero; in the first there is no output rotation, in the second no output torque. At intermediate turbine speeds the efficiency rises to some 80–90 per cent depending on design and operating conditions. The low-turbine-speed inefficiency, with input energy heating the fluid, can be tolerated because these conditions are intermittent, as when starting or hill climbing. The high-turbine-speed inefficiency is avoided by mounting the stator so that it can freewheel in the direction of converter rotation. This occurs as the torque ratio approaches unity and the unit then functions as a fluid coupling, with a maximum efficiency of some 95 per cent (see Figure 18.4).

Since the torque converter provides an infinitely variable torque ratio, up to about 2.5:1 for a single-stage three-element unit, gearbox ratio requirements are simplified. In addition the converter acts as a fluid flywheel, providing an automatic clutch and a torsional damper in the transmission.

Some converters provide torque ratios up to 7:1 by using additional reactor members to redirect the fluid path and impart additional

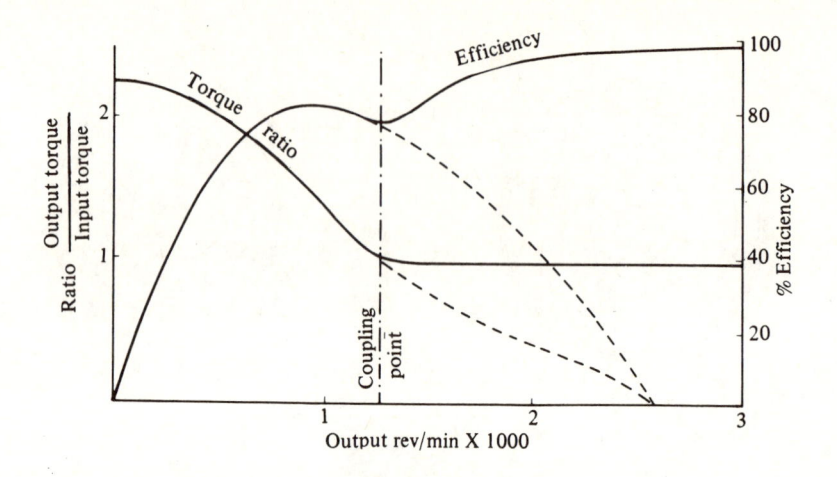

Figure 18.4 Torque-ratio/efficiency curves

impulses to the turbine. In some designs 100 per cent efficiency at higher turbine speed is obtained by locking the turbine to the impeller, and this can be combined with a freewheeling stator.

18.3 Epicyclic gearing

Epicyclic gearing involves one member of the train both revolving on its own axis and rotating bodily about another axis. In a typical application three planet wheels supported on a carrier are in mesh with an inner sun pinion and an outer annulus or internally toothed ring. The planet wheels can revolve on the spindles of the carrier and the planet system rotate bodily around the sun pinion (Figure 18.5).

Sun, planet system and annulus are the three members of an epicyclic gear train which can form one assembly in an epicyclic gearbox. If all three members are free, no drive is transmitted, providing neutral. If any two members are locked together, usually by a hydraulically operated multiplate clutch, the epicyclic train revolves as a unit, giving a 1:1 ratio. If any one member of the train is held stationary, usually by a hydraulically contracted band around a drum carried by that member, gear ratios will be produced between the other two members.

From this typical train a number of alternative gear ratios are available. In an epicyclic gearbox the range of gear ratios is normally obtained by compounding two or more epicyclic assemblies – the output from one train being applied to one member of a second.

For the train discussed, let S be the sun pinion with number of teeth s,

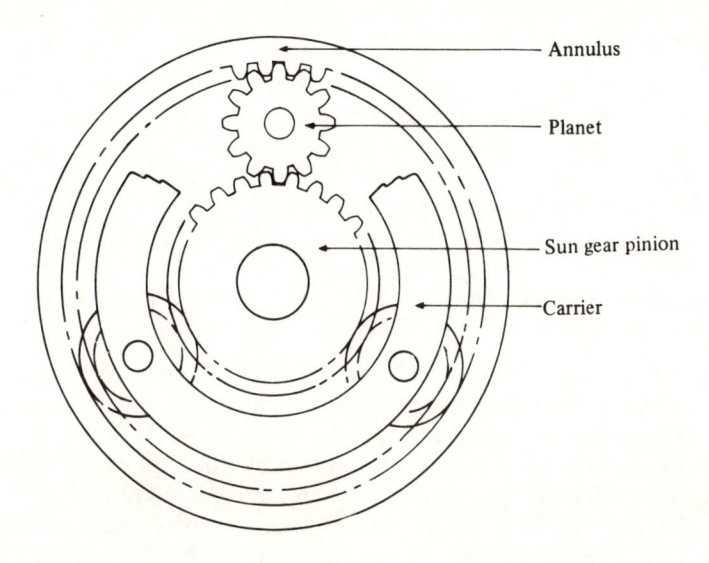

Figure 18.5 Epicyclic gear train

A be the annulus with number of teeth a, and P be the planet carrier assembly. Then

$$\text{teeth on planet gear} = \frac{a-s}{2}$$

The range of gear ratios possible is shown in Table 18.1.

Table 18.1 Epicyclic gear ratios

Member held	Driving member	Driven member	Driving/driven ratio	Driving/driven ratio for a/s ratio		
				2:1	3:1	4:1
P	S	A	a/s*	2:1*	3:1*	4:1*
A	S	P	$(a+s)/s$	3:1	4:1	5:1
S	A	P	$(a+s)/a$	1.5:1	1.33:1	1.25:1

* Driven member turns in reverse direction to driver.
Reciprocal gear ratios can be obtained in each case by making the driver the driven member.

18.4 Automatic transmission

Automatic transmission provides two-pedal control with considerable reduction in driving effort. However, there is also some loss in transmission efficiency compared with a manual clutch and gearbox.

The usual car system uses a torque converter with an epicyclic gearbox. Oil pumps driven by the impeller and the gearbox output shaft operate the brake bands and multiplate clutches through a hydraulic valve block. The control system senses road speed from an output-shaft-driven governor and engine load from the accelerator position.

The system must usually fulfil several requirements:

1. Produce upward gear changes as the road speed rises, but dependent on accelerator position so that these changes occur at lower road speeds under light load
2. Ensure that the change-up road speeds during acceleration are higher than the change-down speeds during deceleration
3. Retain a low gear when required, e.g. when descending a hill and using engine braking
4. Change down at high road speeds when the accelerator is fully depressed – 'kick-down'
5. Prevent the accidental and dangerous engagement of low gear, reverse gear or a parking lock at inappropriate road speeds
6. Prevent starter operation except when the gearbox is in neutral or with the parking lock applied.

Maintenance

The system requires a fluid check and topping-up every 5000 km, and should be drained and refilled at some 40 000 km intervals. As with all hydraulic systems, cleanliness is essential; any particle of cloth or grit will obstruct valve operation.

Owing to overheating problems the vehicle must not be operated at full power for more than a few seconds with the converter turbine stalled. Strict precautions are necessary when power testing a vehicle with automatic transmission in the workshop to preclude the possibility of an accidental gear engagement setting it in motion.

Chapter 19

Universal joints: propeller and drive shafts

A universal joint is a mechanical connection between two shafts with interconnecting axes. It provides a positive drive whilst allowing angular movement of one or both of the shafts.

19.1 Hooke's joint

The Hooke's type of joint or cross arrangement is widely used, and consists of a four-legged spider or cross fitting into Y-shaped yokes on each shaft, needle rollers normally being used to reduce friction (Figure 19.1).

These joints do not transmit the drive at uniform speed when deflected. During each revolution the driven shaft is subject to two accelerations and two decelerations whose values depend upon the angle of deflection – usually limited to about 20°.

When an intermediate shaft, e.g. a propeller shaft, has Hooke's joints at each end, provided the yokes and the angles of deflection at each are equal, the cyclic variation of the intermediate shaft will be cancelled out by the second joint. The speed fluctuation of the intermediate shaft remains, and this has resulted in the replacement of the Hooke's joints by constant-velocity joints in many applications.

Double, back-to-back, Hooke's joints have been used as constant-velocity joints, but are now replaced by more compact devices.

19.2 Constant-velocity joints

One principle for constant velocity (CV) depends upon the engagement between the two shafts taking place in the plane bisecting the angle between them.

In the most common – Rzeppa – type of CV joint, caged ball bearings transmit the torque between longitudinal grooves in the end members; these have a cross-section to load the balls in compression rather than in shear (Figure 19.2). In the earlier types the balls were track steered, by

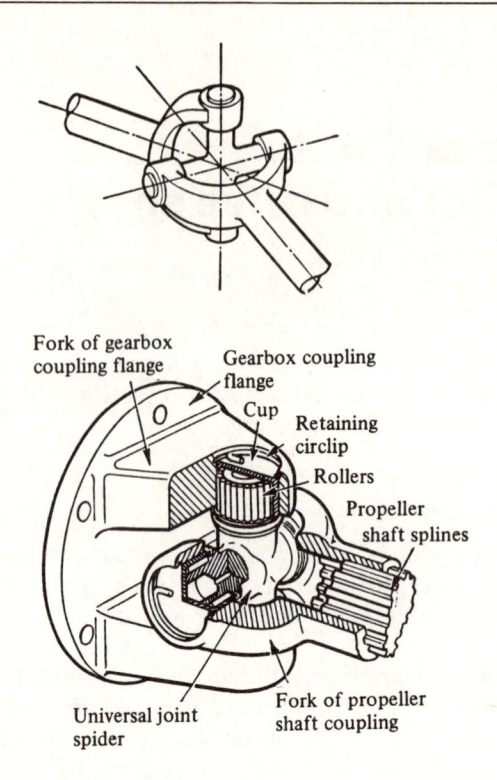

Figure 19.1 Hooke's universal joint

the geometry of the curvature of the grooves, to lie in the bisecting plane. A later development uses the geometry of the inner and outer surfaces of the cage to steer the balls, and this permits straight instead of curved ball grooves. The inner member can now move bodily along the bore of the outer member and provide the plunge action required for suspension movements. These plunge-type CV units are widely used to eliminate the sliding-splined joint previously required (Figure 19.3).

In addition to providing constant velocity these joints will operate at approximately twice the angle of divergence (40–45°) of a Hooke's joint, and are particularly used for the outboard couplings on front-wheel drive.

19.3 Flexible joints

In the earlier type of flexible-ring joint, a disc of reinforced rubberized fabric was bolted between three-armed spiders carried on the shafts

Universal joints: propeller and drive shafts

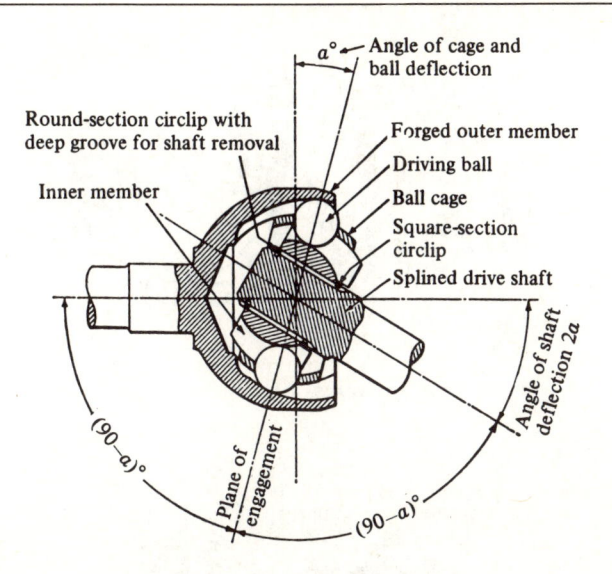

Figure 19.2 Rzeppa constant-velocity joint

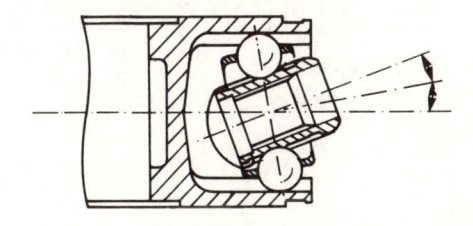

Figure 19.3 Plunge constant-velocity joint

(Figure 19.4). In the modern resilient coupling the angular deflection (some 5° for continuous running and 15° maximum) is obtained by the elastic deformation of rubber elements. The rubber, of a mix to give the required characteristics, is pre-compressed to ensure maximum life and torque capacity (Figures 19.5, 19.6).

In the usual type, spherical rubber cups or bushes are bonded to a metal ring and to metal attachment sleeves for the spiders on the drive shafts. Alternatively a complete rubber ring with bonded-in sleeves may be used. For the inboard couplings on front-wheel drive, the torsional flexibility of these joints can cushion shock loading and reduce vibration in an otherwise short rigid transmission. Their axial flexibility is often adequate to eliminate sliding splined joints.

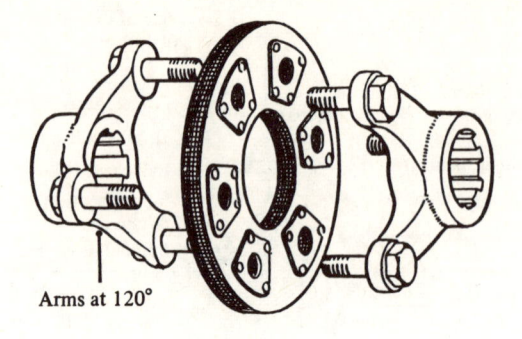

Figure 19.4   Flexible-ring universal joint

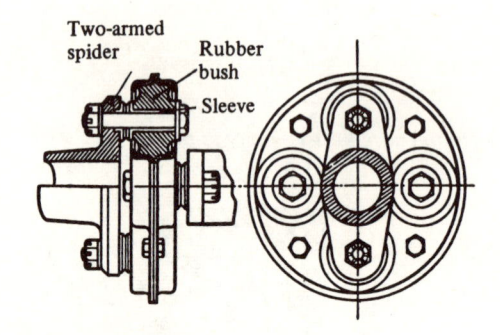

Figure 19.5   Layrub universal joint

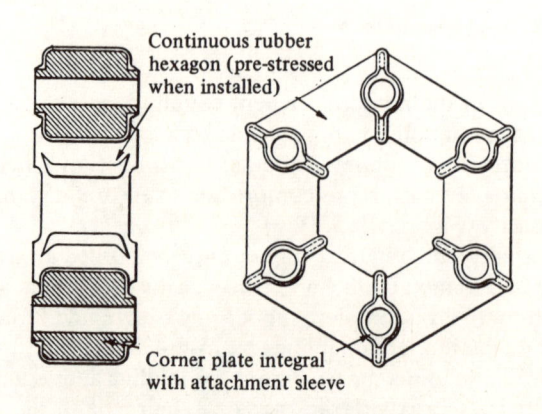

Figure 19.6   Rotoflex flexible-ring univeral joint

19.4 The propeller shaft

The propeller shaft connects the gearbox to a live rear axle or a sprung rear drive unit. It revolves at engine speed in direct drive, and accurate dynamic balance is essential to prevent vibration.

The critical whirling speed where resonant vibration will occur is inversely proportional to the square of the length. For example, if the maximum speed for a 1 m shaft is 5500 rev/min, that for a 1.25 m shaft of the same type is only 3500 rev/min.

In order to obviate vibration and noise the shaft length may be reduced by a rearward extension on the gearbox or a forward extension on the rear drive unit. Alternatively a divided propeller shaft is often used. The 'primary' shaft connects to the gearbox mainshaft through the universal joint and has a rear support consisting of a sealed deep-groove ball bearing in a housing, rubber mounted to insulate vibration from the body and accommodate any slight distortion. The 'secondary' shaft has two universal joints and a sliding-splined joint, usually at the front end.

For commercial vehicles with a wheelbase exceeding about 5 m a three-piece propeller shaft, with two intermediate support bearings, four universal joints and a sliding joint will be needed.

19.5 Drive shafts

The drive shafts connect to independently sprung front or rear road wheels. Being on the output side of the final-drive unit, they operate with that ratio of torque increase and speed reduction compared with a propeller shaft. If inboard-mounted brakes are fitted, the shafts and universal joints must also withstand braking torque – often considerably greater than driving torque.

In most cases the shafts are of equal length and typically of 20–25 mm diameter nickel-chromium steel. With a transversely mounted engine-transaxle unit, unequal-length shafts may be required. The longer shaft can be partly of larger-diameter tubular construction to equalize the torsional rigidity and vibration characteristic with the shorter solid shaft.

19.6 Maintenance

For vibrationless running, propeller shafts and (to a lesser extent, since they revolve at lower rev/min) drive shafts must be dynamically balanced and run true. Any wear on universal joints or sliding splines which can cause misalignment will produce out-of-balance forces and cause vibration.

Hooke's joints are frequently replaced as a propeller shaft assembly for cars, or alternatively by fitting new spider, needle rollers and cups. They are usually assembly packed with lithium grease but some also have provision for periodic greasing.

The constant-velocity joints, being a type of ball race, are very vulnerable to dirt and water entering through faulty rubber gaiters. They are normally replaced as a unit.

The Rotoflex flexible-ring joint is pre-stressed by an encircling steel band, which is removed after assembly and requires to be similarly compressed by a suitable tool before removal.

Chapter 20
The final drive and differential

The final-drive gears may be directly driven from the gearbox and incorporated in a front-mounted engine-transaxle unit. Alternatively they may be mounted in a live rear axle or in a sprung rear drive unit.

20.1 Bevel pinion and crown wheel

The bevel pinion is carried by taper-roller bearings supporting radial and axial loads. A shim adjustment allows the depth of engagement with the crown wheel to be varied, and shims or a collapsible spacer provide for the specified pinion-bearing pre-load torque.

The crown wheel, bolted to the differential assembly, is also supported in taper-roller bearings and has lateral adjustment by shims or screwed rings. Commercial final-drive units frequently employ a straddle pinion mounting with an additional parallel-roller bearing located on the inner side of the gear teeth, and also a kick-pad adjacent to the rear face of the crown wheel to minimize deflection at the point of engagement.

The straight-toothed bevel gear has been replaced by the quieter-running spiral (helical) bevel, where the tooth curvature provides a greater contact area and spreads the load beyond a single pair of teeth (Figure 20.1).

Although the advantage of a lower propeller shaft for cars and some commercial vehicles has caused the replacement of the spiral bevel by the hypoid drive in live axles, it is frequently used in transaxle units.

20.2 Hypoid gear

In this drive the pinion is offset from the crown-wheel's axis, in its direction of rotation, by up to some 20 per cent of the diameter (Figure 20.1). The pinion size, tooth strength and spiral angle are increased, with benefits in durability and silence, whilst it provides for a lower propeller shaft.

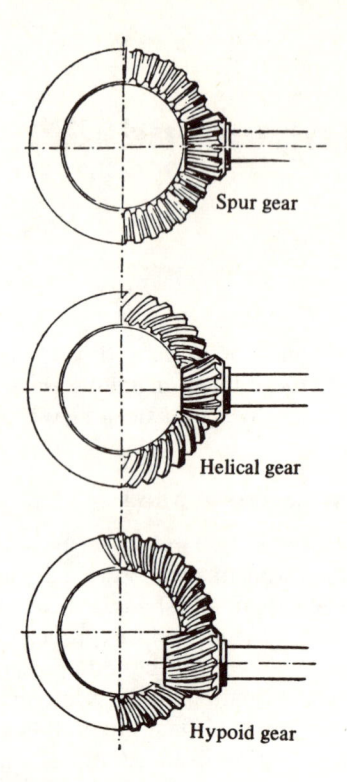

Figure 20.1 Bevel final-drive gears

The sliding action of the teeth necessitates an extreme-pressure lubricant; an oil satisfactory for spiral-bevel gears can result in early tooth failure.

The crown wheel runs with the lower teeth submerged in lubricant which is distributed around the casing; a catch trough may be used to direct the flow to the pinion and bearings.

20.3 Worm and wheel

For heavy vehicles, worm drive offers silence, durability, a large gear reduction and an easy facility for driving a second axle. It is costly and, with a purely sliding action, has a somewhat lower mechanical efficiency than bevel gearing.

A steel worm engages, from above or below, with a phosphor-bronze worm wheel carried by the differential gear assembly. The worm may be parallel or hour-glass – conforming to wheel curvature to eliminate the heavy end-thrust (Figure 20.2).

The worm is multistart and the gear ratio is the number of teeth on the wheel divided by the number of starts on the worm. Special lubricants are required to accommodate the sliding action of the teeth.

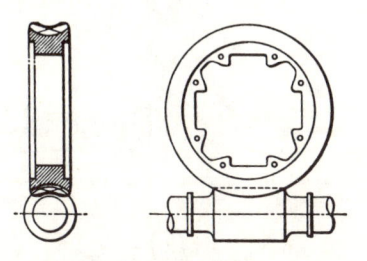

Figure 20.2   Worm-and-wheel final-drive gear

20.4   Two-speed epicyclic final drive

An alternative to providing additional ratios at the gearbox is the use of the two-speed epicyclic final-drive unit. The usual arrangement consists of an annulus formed on the inside of the crown wheel which drives four planet pinions mounted on a carrier formed as an extension of the differential cage. The hollow sun pinion can slide one way and lock to the planet carrier, so locking the epicyclic train and giving high or direct drive to the differential cage. Slid the other way – by pneumatic or electrical operation – the sun is locked to the axle casing and the planet wheels revolving around the fixed sun carry the planet carrier and the differential cage at some 70–75 per cent of the crown-wheel rev/min. It will be noted that the action is similar to that of the epicyclic overdrive unit but with the drive passing the opposite way through the gear train.

20.5   The differential

The differential provides an equal torque to each half-shaft or drive shaft although they may be rotating at different speeds. It therefore allows the outer road wheel to revolve faster than the inner when cornering, whilst maintaining a positive drive to both wheels.

The differential cage, to which is bolted the crown wheel or worm wheel, contains two central bevel or sun wheels which are splined internally for the half- or drive shafts (Figure 20.3). Meshing with each sun wheel are two planet pinions carried by and free to revolve on a

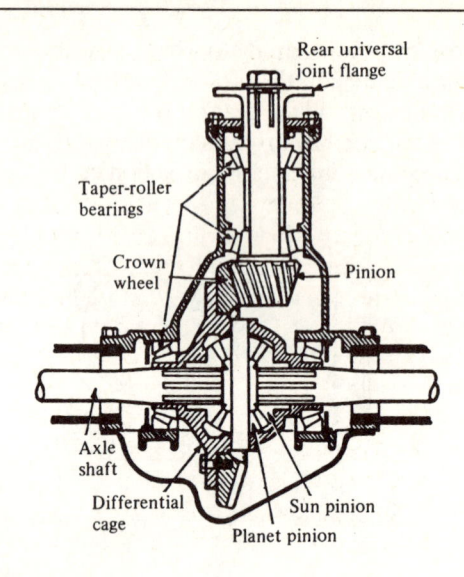

Figure 20.3   Rear-drive unit

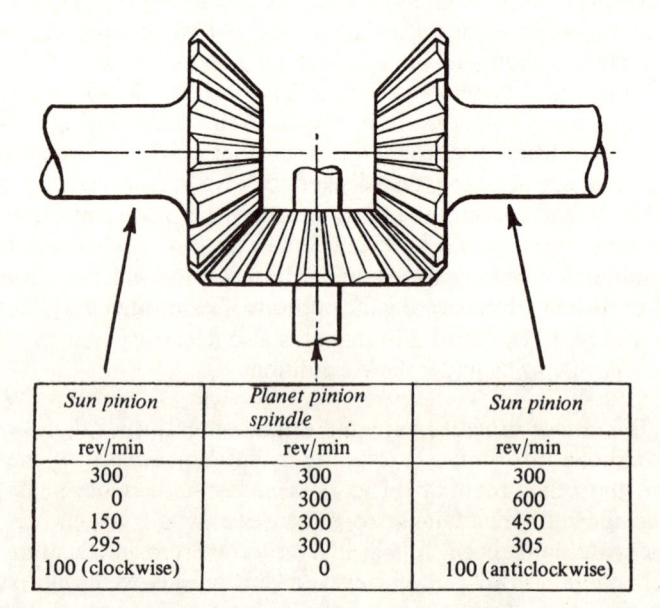

Sun pinion	Planet pinion spindle	Sun pinion
rev/min	rev/min	rev/min
300	300	300
0	300	600
150	300	450
295	300	305
100 (clockwise)	0	100 (anticlockwise)

Figure 20.4   Differential action

spindle fixed in the differential cage. Commercial vehicles usually employ a differential-cage construction in two halves, carrying a spider with four planet wheels.

The planet wheels correspond to a balance beam and provide an equal driving force on each sun pinion of half the force applied to the spindle or spider by the differential cage (Figure 20.4).

During normal straight-ahead running the planet pinions carry each sun pinion around at the same speed as the differential cage. If the differential cage rotates when one drive shaft is held stationary and the other is free to revolve, the planet pinions must roll around the stationary sun wheel and drive the opposite sun wheel at twice the speed of the differential casing.

Normal cornering falls between these two conditions, and a reduction in the speed of one sun pinion compared with the crown-wheel speed is balanced by a corresponding increase in the speed of the other sun pinion. For example, with a crown-wheel speed of 300 rev/min, if the inner shaft revolves at 295 rev/min the outer shaft will travel at 305 rev/min.

Since no greater torque can be transmitted to one road wheel than the other owing to the balance action of the planet pinions, if one road wheel is on a slippery surface where it can revolve idly, no tractive force can be applied to the other wheel. It follows that if the differential cage is held stationary and one road wheel is rotated, the other wheel must revolve at an equal speed in the opposite direction.

20.6 Differential applications

When a vehicle is cornering the outside front wheel turns fastest and the inside rear the slowest; the combined turns of the front wheels exceed those of the rear. When four-wheel drive is used on hard ground, a third differential is necessary to balance the torque and prevent a torsional wind-up in the transmission due to the variation of rotation between front and rear. The third differential is also necessary between driven twin axles operating under these conditions.

Vehicles using a two-speed transfer gearbox for optional 4WD for cross-country use omit the third differential as it is both unnecessary and disadvantageous on loose ground.

The limitations of the differential – loss of drive if one wheel spins – can be eliminated by the use of a differential lock, e.g. dog clutching one of the sun pinions to the differential cage. Alternatively, a limited-slip arrangement with solid or viscous friction between the sun wheel and the differential cage will apply some torque to the wheel having grip.

Off-the-road vehicles sometimes drive the wheels through ratchets, so allowing the outer wheel to over-run when turning. This avoids wheel spin but does not equalize the driving torque.

20.7 Maintenance

Both crown wheel and pinion must be adjusted axially to obtain the correct tooth contact with the load evenly distributed over the side of the tooth. The backlash between the teeth varies with the design and is usually some 0.12 mm to 0.25 mm (Figure 20.5).

In most cases the bearings are pre-loaded to reduce the deflection of the gears caused by their tendency to separate when under load. Typical figures for steady rotation (i.e. avoiding starting torque) are 1–2 Nm for the pinion without the oil seal fitted and 2–3 Nm for the drive gear. On most car designs these pre-loads and the correct meshing are obtained by shim adjustment. In many cases, special fixtures are required to establish the datum position of the pinion, and sometimes the casing must be spread slightly to fit and remove the differential assembly. A typical procedure would be:

1. Determine the shims necessary to give the required pre-load on the differential assembly (e.g. 0.12 mm total pre-load). Remove differential assembly from the axle housing.
2. Install and adjust the pinion to the specified height using the manufacturer's gauge.
3. Adjust the pinion bearings to give the correct pre-load.
4. Install the drive-gear assembly and determine the backlash between the teeth. Remove and apportion the selected shims on each side to give the correct backlash.
5. Using marking blue, examine the driven-face tooth contact for about a dozen teeth on the crown wheel. If this is satisfactory the over-run side contact is normally correct.
6. Make any necessary correction, whilst maintaining the correct backlash and pre-loads. In general, related movements of both pinion and drive gear are necessary to maintain the correct backlash.

If the mesh of the gears is not deep enough there is usually a noise when the vehicle over-runs the engine; if the mesh is too deep there will be a whine on engine drive.

Example 20

A car has one rear wheel jacked up, and with top gear engaged this wheel turns three times while the engine is cranked seven turns. With low gear engaged, 17 turns of the engine rotate the rear wheel twice. Calculate the rear-axle ratio and gearbox ratio on low gear.

The final drive and differential

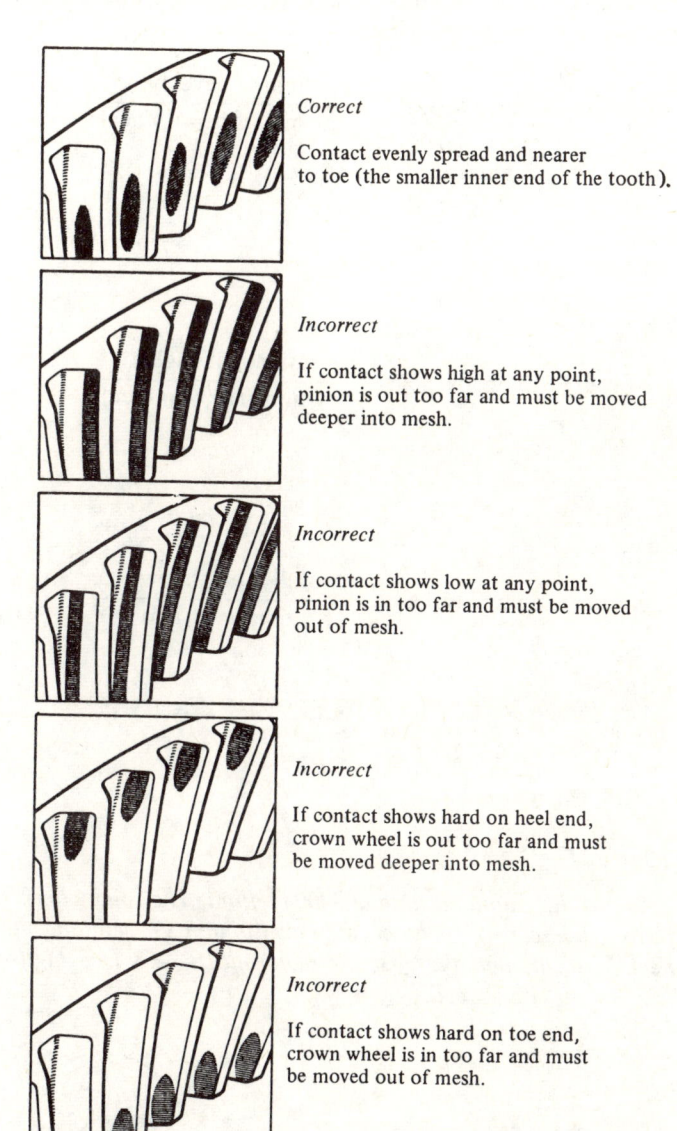

Correct

Contact evenly spread and nearer to toe (the smaller inner end of the tooth).

Incorrect

If contact shows high at any point, pinion is out too far and must be moved deeper into mesh.

Incorrect

If contact shows low at any point, pinion is in too far and must be moved out of mesh.

Incorrect

If contact shows hard on heel end, crown wheel is out too far and must be moved deeper into mesh.

Incorrect

If contact shows hard on toe end, crown wheel is in too far and must be moved out of mesh.

Figure 20.5   Crown-wheel tooth contact

Since only one rear wheel is turning, it will make twice the number of turns of the crown wheel. Hence:

$$\text{rear-axle ratio} = \frac{\text{rev/min of bevel pinion}}{\text{rev/min of crown wheel}}$$

$$= \frac{7}{1.5}$$

$$= 4.67$$

Now

$$\text{overall ratio on low gear} = \frac{\text{rev/min of engine}}{\text{rev/min of crown wheel}}$$

$$= \frac{17}{1}$$

Hence

$$\text{gearbox ratio on low gear} = \frac{\text{overall ratio on low gear}}{\text{rear-axle ratio}}$$

$$= \frac{17}{4.67}$$

$$= 3.64$$

Example 21

A van has the engine turning at 3000 rev/min. The road wheels have an effective diameter of 0.868 m, the bevel pinion 17 teeth and the crown wheel 72 teeth, and the gearbox rear ratio is 3.58 (see Figure 20.6). Calculate the road speed.

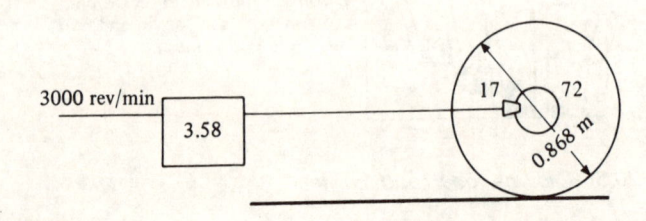

Figure 20.6

The final drive and differential

We have

$$\text{gearbox gear ratio} = 3.58$$

$$\text{rear-axle ratio} = \frac{72}{17}$$

$$\text{overall ratio} = 3.58 \times \frac{72}{17}$$

$$\text{rev/min of engine} = 3000$$

Thus

$$\text{rev/min of road wheels} = 3000 \times \frac{17}{3.58 \times 72}$$

$$\text{distance covered per hour} = \text{rev/min} \times \text{circumference in m} \times \frac{60}{1000} \quad \text{km}$$

$$= 3000 \times \frac{17}{3.58 \times 72} \times \pi \times 0.868 \times \frac{60}{1000} \quad \text{km}$$

Hence

$$= 32.4 \quad \text{km}$$

$$\text{speed of vehicle} = 32.4 \quad \text{km/h}$$

Example 22

An engine develops a torque of 118 Nm at the flywheel at 2500 rev/min and drives through constant-mesh pinions of 17 teeth on the clutch shaft and 30 on the layshaft (Figure 20.7). Second-gear pinions on layshaft and mainshaft have 20 and 27 teeth respectively, and the bevel pinion and crown wheel 11 and 50 teeth. Calculate the speed and torque at each axle shaft on top and second gear, presuming equal speed of each shaft.

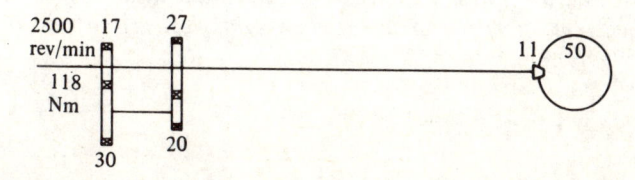

Figure 20.7

Top gear

$$\text{rear-axle ratio} = \frac{50}{11}$$

$$\text{torque at each rear-axle shaft} = \frac{118}{2} \times \frac{50}{11} \text{ Nm}$$

$$= 268.18 \text{ Nm}$$

$$\text{speed of axle shafts} = 2500 \times \frac{11}{50} \text{ rev/min}$$

$$= 550 \text{ rev/min}$$

Second gear

$$\text{gearbox ratio} = \frac{27}{20} \times \frac{30}{17}$$

$$= 2.38$$

$$\text{torque at each axle shaft} = 268.18 \times 2.38 \text{ Nm}$$

$$= 638.27 \text{ Nm}$$

$$\text{speed of axle shafts} = \frac{550}{2.38} \text{ rev/min}$$

$$= 231 \text{ rev/min}$$

Example 23

A car has a mass of 1200 kg. The engine develops a maximum torque of 140 Nm and the rear-axle ratio is 5.09:1. The rear wheels are 0.66 m in diameter and the coefficient of friction between the tyres and the road is 0.7. Calculate the lowest useful gearbox ratio under these conditions if the rear wheels carry 60 per cent of the total weight. Neglect transmission loss. ($g = 9.8$ m/s^2.)

We have

$$\text{weight on rear wheels} = \frac{60}{100} \times 1200 \times 9.8 \text{ N}$$

The final drive and differential

max. tractive force at rear wheels without wheel spin

$$= \mu \times \text{normal force on wheels}$$

$$= 0.7 \times \frac{60}{100} \times 1200 \times 9.8 \quad \text{N}$$

max. useful rear-axle torque = max. tractive force × radius of tyre

$$= 0.7 \times \frac{60}{100} \times 1200 \times 9.8 \times \frac{0.66}{2} \quad \text{Nm}$$

$$= 1630 \quad \text{Nm}$$

Now

$$\text{max. engine torque} = 140 \, \text{Nm}$$

Thus

$$\text{max. useful overall gear ratio} = \frac{1630}{140}$$

$$\text{max. useful gearbox ratio} = \frac{1630}{140} \times \frac{1}{5.09}$$

$$= 2.29$$

Chapter 21
The rear-wheel assembly

21.1 The live rear axle

The live-rear-axle assembly has to act as a bearing for the road wheels and provide attachments for the springs. It transmits the drive from the propeller shaft to the road wheels and, serving as a beam subject to a bending load, supports the weight of the rear of the vehicle.

By means of the final-drive unit the drive is turned at right angles, and a gear reduction is obtained of about 3.5:1 to 5.0:1 for cars (greater for commercial vehicles). This rev/min reduction gives the necessary torque required for direct drive.

21.2 Axle casing

The most usual is the carrier axle, where a malleable cast-iron centre section provides rigidity for the final-drive gears, while a bolted-on rear cover allows assembly and removal. Tubular-steel axle tubes are pressed in and welded; the carrier section can thus accommodate various widths of wheel track (Figure 21.1).

The banjo construction is welded up from steel pressings, including the rear cover and the round or square arms. The final-drive assembly bolts in from the front.

21.3 Axle-hub construction

The inner ends of the half-shafts are splined to fit the differential sun pinions. The methods of supporting the outer ends are usually divided into three classes (semi, three-quarter and fully floating) according to the method of bearing mounting.

Semi-floating rear hub

The usual arrangement is for the nickel-chromium steel half-shaft to have a substantially increased diameter at the outer end and a forged

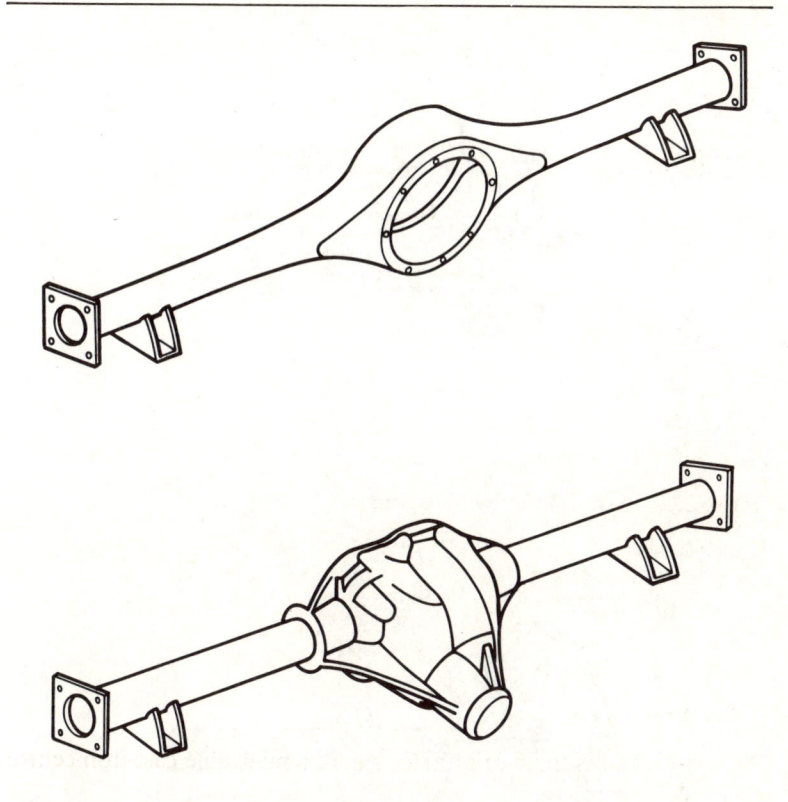

Figure 21.1   Carrier and banjo axle casings

integral flange to carry the brake drum and wheel studs. A deep-groove ball bearing, often grease-sealed, is retained on the shaft by an interference-fit ring and in the enlarged end of the axle casing by a bolted-on plate which also secures the brake back plate (Figure 21.2).

A single taper-roller bearing is sometimes used with a keyed taper fit for the hub. In this case a spacer is fitted between the inner ends of the axle shafts so whilst outward thrust on either wheel is taken by the adjacent bearing, inward thrust is transferred to the opposite one; shim adjustment limits the end-float to some 0.1 mm on assembly.

The half-shaft must support:

(a) Shear loading due to the vehicle's weight
(b) Bending loads due to the vehicle's weight, and cornering, braking and accelerating forces
(c) Driving torque.

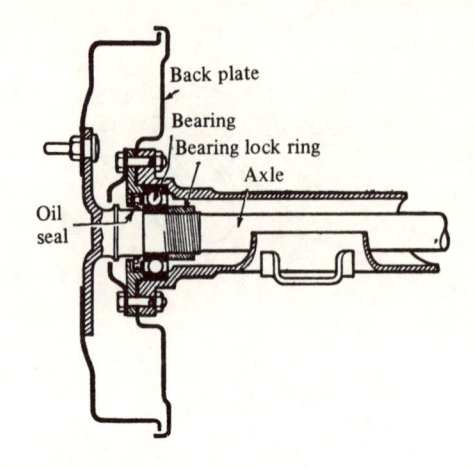

Figure 21.2 Semi-floating hub layout

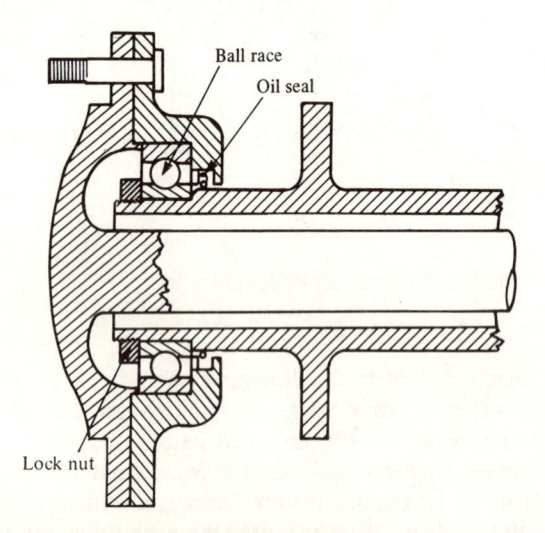

Figure 21.3 Three-quarter floating hub layout

Three-quarter-floating rear hub

This less frequently employed design uses a large-diameter single- or double-row ball race between the hub and the outside of the axle casing. The axle shaft is thus relieved of shear and bending loads due to the vehicle's weight; there remains some bending due to cornering forces, and also the torque loading (Figure 21.3).

Fully floating rear hub

The half-shaft transmits driving torque only and its removal or failure does not affect the road wheel, which is supported by widely spaced double taper-roller bearings on the outside of the axle casing. The system is generally used with heavy vehicles and also for dead axles (Figure 21.4).

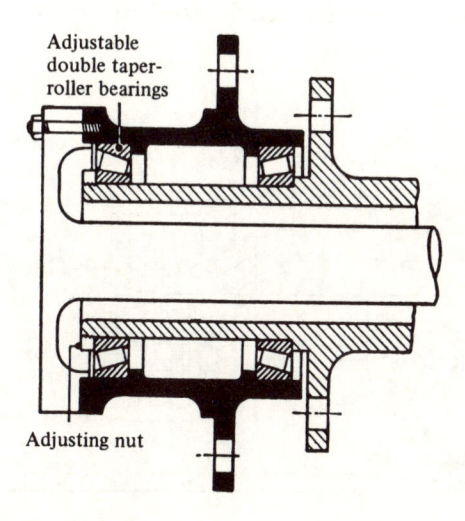

Figure 21.4 Fully floating hub layout

Oil sealing and venting

The hub bearings may be grease-sealed or greased, or may rely on rear-axle lubricant. Lip-type seals and O-rings are used where applicable to contain the lubricants. Axle oil can attain high temperatures, and the accompanying pressure rise must be vented from the axle casing through a breather.

21.4 The sprung rear-drive unit

The gear arrangements in a separately mounted rear-drive unit, for use with independent rear suspension, are the same as for a live rear axle. The axle shafts are replaced by stub drive shafts supported and located by deep-groove outer ball races, with lip-type oil seals – in effect a 'shortened' semi-floating construction.

21.5 The transaxle unit

Occasionally the gearbox and rear-drive unit are combined into a sprung rear-mounted transaxle. More usually the transaxle is part of the engine-transmission assembly for front-wheel-drive vehicles (Figures 21.5, 21.6).

In the transaxle casing, the bevel pinion – integral with the gearbox output shaft – is usually supported by double taper-roller bearings in the gearbox portion, and projects into the axle part.

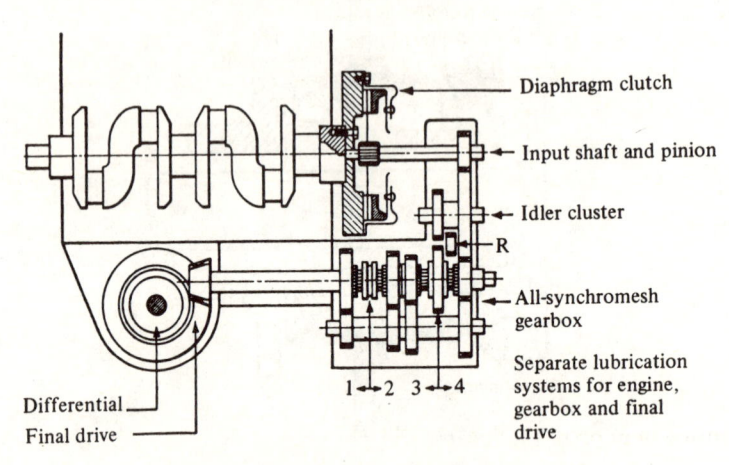

Figure 21.5 Transaxle unit: longitudinal engine

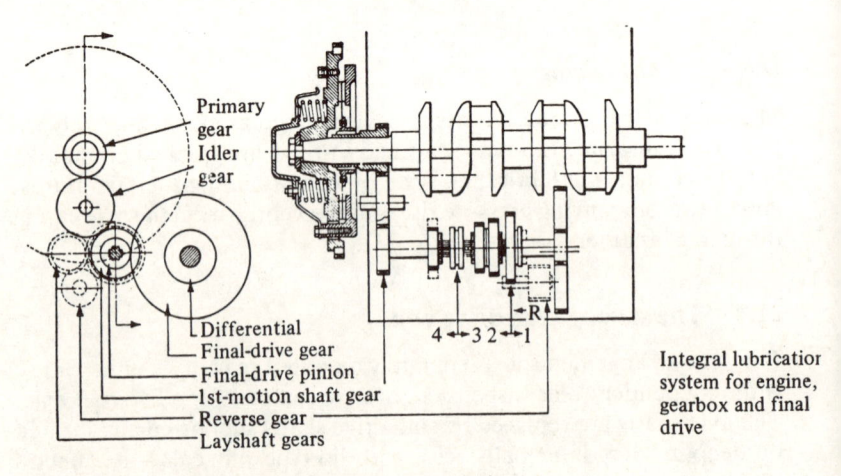

Figure 21.6 Transaxle unit: transverse engine

21.6 Torque reaction and driving thrust

With a live rear axle the torque reaction on the axle casing must be resisted to apply the equal and opposite driving torque to the road wheels. This driving torque acting at the area of contact of the tyre and the road creates the tractive effort propelling the vehicle, and this again results in a driving thrust on the axle casing which must be transmitted to the vehicle structure. When the brakes are applied the direction of these forces is reversed.

There is also a moment produced by the bevel pinion torque that tends to rotate the axle casing about the pinion axis in the direction of rotation and so lift the right-hand road wheel.

Originally two systems of live-axle control were used – Hotchkiss and torque-tube drive.

Hotchkiss drive

In the Hotchkiss drive, two semi-elliptic springs are fitted, being shackled at the rear and transmitting the driving thrust through pivot pins at the front ends (Figure 21.7).

Driving-torque reaction and braking torque are resisted by the springs, and during acceleration and deceleration considerable twisting of the axle casing and flexure of the springs takes place.

For this reason, and because the propeller shaft and the axle movement do not follow the same arcs when the springs are deflected, it is necessary to have a second universal joint at the rear of the propeller shaft and a splined joint to allow for the variation of length between centres. This sliding joint may be at the front of the propeller shaft or at the rear end of the gearbox mainshaft.

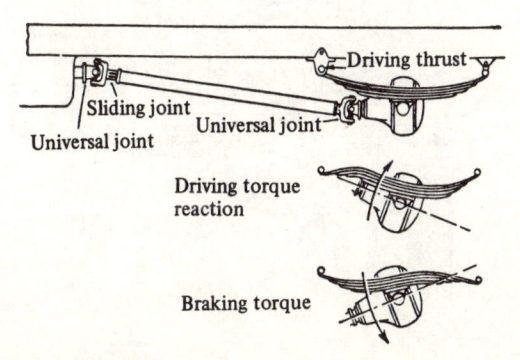

Figure 21.7 Hotchkiss drive

Torque-tube drive

The torque tube, surrounding the propeller shaft, is rigidly bolted to the axle casing at the rear and housed in a spherical joint at the front end. This spherical bearing takes the driving thrust, driving-torque reaction and braking torque. The propeller shaft is driven through a universal joint situated at the centre of the spherical bearing (Figure 21.8).

The two semi-elliptic springs, shackled at both ends to permit the arcuate axle movement and bolted to pivoting saddles to eliminate twist, can have a rate suited only to suspension requirements.

The current semi-torque tube has a much reduced unsprung mass, consisting of a pinion-housing extension tube with a front rubber mounting on to a body cross member (Figure 21.9). The short propeller shaft may have plunge-type CV joints rather than Hooke's joints with a sliding spline. Coil springs are used and a system of axle control is essential.

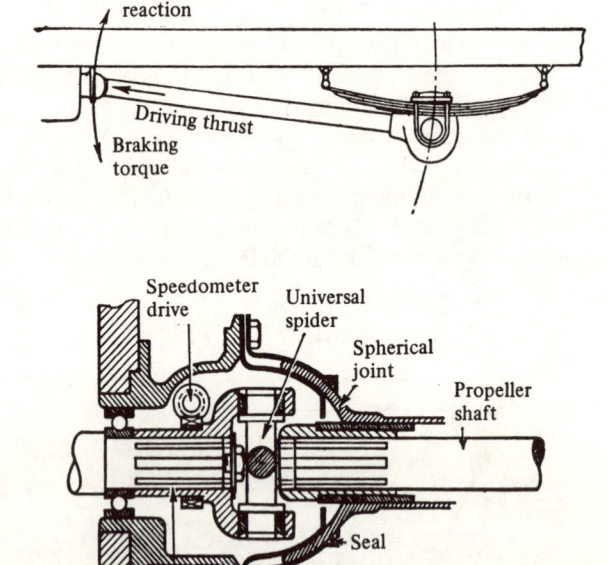

Figure 21.8 Torque-tube drive

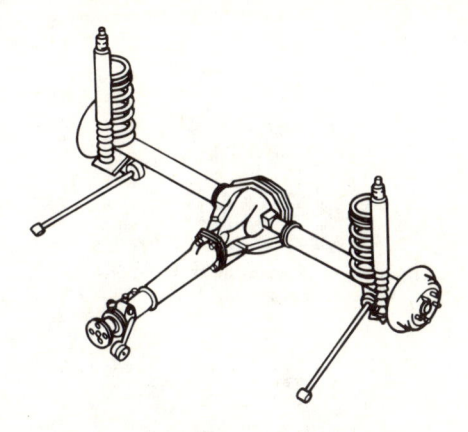

Figure 21.9 Semi-torque-tube drive

21.7 Live-axle control

Lateral control for a live axle using coil springs can be achieved by a simple transverse rod, rubber mounted to the body on one side and to the axle on the other – a Panhard rod. The more elaborate Watt's linkage may sometimes be used, where a vertical link, pivoting on the axle casing, has transverse connecting rods to the body from top and bottom.

In the five-link axle-control system, two trailing arms on each side connect from the body above and below the axle tubes, together with a

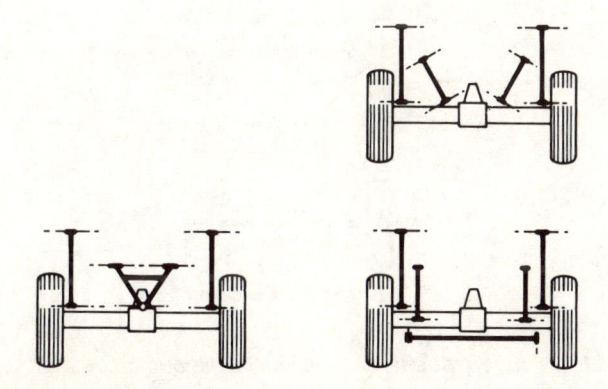

Figure 21.10 Live-axle control: five link; angled upper links; wishbone and twin links

(Plan views)

Pure swing axle (obsolete)

Pure trailing arm

Semi-trailing arm

(End views)

Unequal transverse links

Transverse link and strut

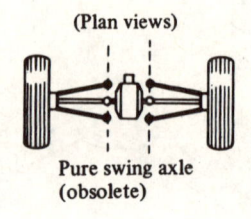

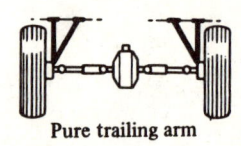

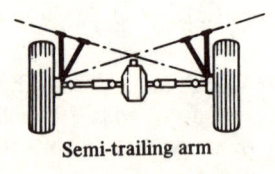

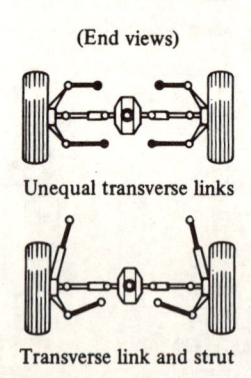

Figure 21.11 Independent rear suspension layouts

Panhard rod. This arrangement can be simplified by angling the two upper links to provide lateral control. Alternatively, the upper links can be replaced by a central wishbone bracket pivoting on the body (Figure 21.10).

A coil-sprung live axle with adequate suspension geometry and control of torque reaction and driving thrust is robust and reasonably acceptable, but independent rear suspension halves the unsprung mass and provides substantial advantages at an increased complexity and cost.

21.8 Independent rear suspension

A typical arrangement is a subframe attached to the vehicle structure through vibration- and sound-insulating rubber mountings. The wheels and brake assemblies are carried on semi-trailing arms, which may be fabricated steel stampings or light-alloy castings. The drive shafts require two universal joints and a sliding joint.

Semi-trailing arms operate on diagonal axes radiating forward from the nosepiece of the final drive, and are thus between the pure trailing arm, with a transverse axis parallel to and ahead of the drive shaft, and the pure swing axle, with a longitudinal axis through the inboard universal joint. The latter (obsolete) system alters camber as the wheel lifts and can introduce unstable cornering (Figures 21.11, 21.12).

Alternatively a transverse link-and-strut system (MacPherson; termed Chapman when used at the rear) or double transverse links may

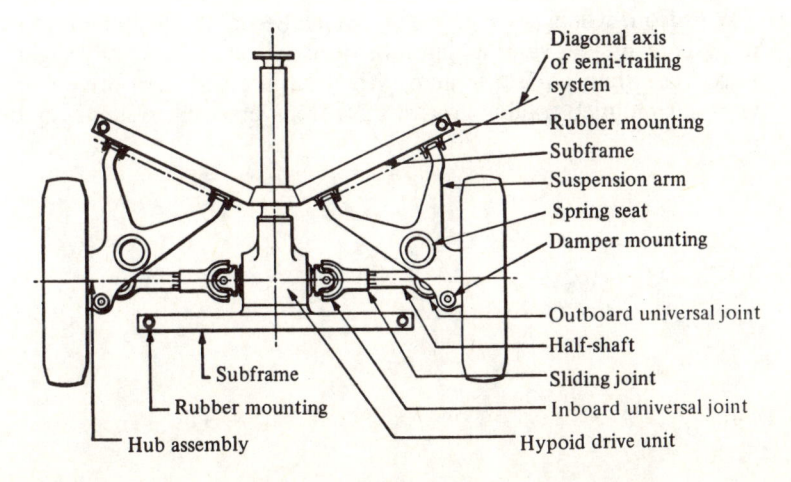

Figure 21.12 Semi-trailing-arm independent rear suspension

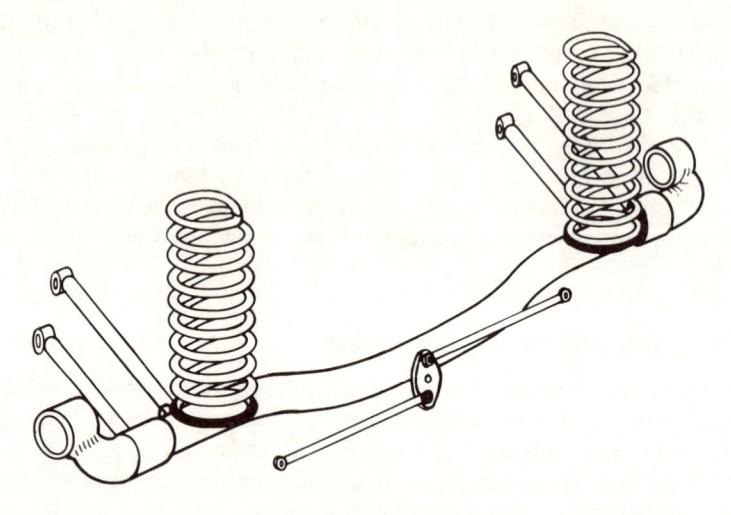

Figure 21.13 De Dion tube with Watt's linkage and twin trailing links

be used, with either a longitudinal radius arm or a wide wishbone to resist the driving and braking forces. On occasion the upper transverse link has been replaced by the drive shaft using two Hooke's joints.

On a few designs, a cranked tube connects the two rear hub carriers and maintains the wheels parallel. This De Dion tube can be located by twin trailing links each side with a Watt's linkage to give accurate lateral control (Figure 21.13).

With front-wheel-drive cars, independent rear suspension has more limited advantages since a light-tube dead axle need have only slightly more mass than an IRS system. When required for non-driven rear wheels, simplified trailing-arm, strut or transverse-link systems can be used.

Chapter 22

The braking system

22.1 Principles

When the brakes are applied on a moving vehicle, the kinetic energy (the energy of motion of the vehicle, given by $(1/2) \times \text{mass} \times \text{velocity}^2$) is transformed into heat energy generated between the braking surfaces. Approximately the same energy dissipation is required to halt a vehicle from 80 km/h as to descend, without increasing speed, one kilometre of road with a gradient of 1:40 in the same vehicle.

Consider a simple drum brake. The force of friction or retarding force between the linings and the drum depends upon the coefficient of friction for the two materials and the force exerted by the shoes on the drum. Thus, from Figure 22.1,

$$F_B = \mu_B W_B$$

where F_B = force of friction or retarding force on the brake drum, μ_B = coefficient of friction (approximately 0.3–0.4 for these surfaces), and W_B = normal force of shoes on drum.

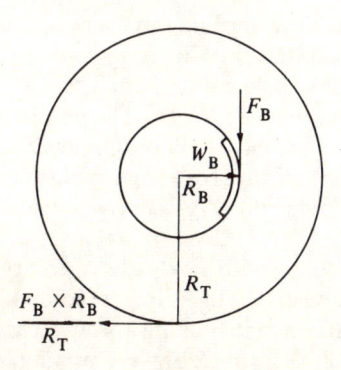

Figure 22.1 Braking forces

The retarding force acting on the brake drum multiplied by the radius of the drum R_B gives the retarding torque acting on the wheel:

$$F_B R_B = \text{retarding torque}$$

Dividing the retarding torque by the radius of the tyre R_T will give the retarding force produced at the area of contact of the tyre on the ground:

$$\frac{F_B R_B}{R_T} = \text{retarding force produced on the ground}$$

Now the maximum retarding force that can be employed depends upon the force of friction between the tyre and the ground, and this in turn depends upon the coefficient of friction for the two surfaces and the normal force on the tyre. Thus:

$$F_T = \mu_T W_T$$

where F_T = force of friction between tyre and ground, μ_T = coefficient of friction (very variable with the conditions; say 0.8), and W_T = normal force on tyre.

In practice the maximum retarding force is obtained just before the wheel locks and skidding occurs. The advantage of sensing systems to detect incipient wheel locking and monitor brake operation thus allows maximum retardation as well as preserving control. It seems likely that the electronic facility will provide the most effective and economical management system.

22.2 Weight transfer

During deceleration the retarding force acts at ground level, whilst the inertia of the car, produced by its mass and speed, acts through the centre of mass in the opposite direction. These two equal and opposite forces produce an overturning couple. The perpendicular force between the wheels and the ground is therefore increased at the front and decreased at the rear to form a resisting couple during deceleration. The amount of accompanying body *brake dive* is determined by the suspension layout (Figure 22.2).

Cars, especially those with FWD, have a large proportion of the weight on the front wheels. This is increased by weight transfer during braking, so that large front braking torques can be applied without skidding. Equally some form of brake limitation may be necessary to prevent the rear wheels sliding.

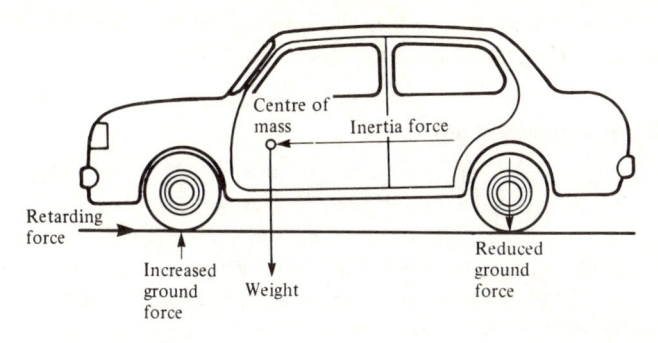

Figure 22.2 Weight transfer

22.3 Brake shoes and drum

Asbestos, a fibrous mineral with good frictional and heat-resisting properties, is the basis of most brake and clutch linings. Asbestos yarn or cloth – sometimes incorporating brass or zinc wire – is impregnated with resinous bonding liquids and subjected to pressure and heat to form the wound or woven materials. For the moulded type of liner, which can offer a greater range of characteristics, the random fibres are combined with a variety of synthetic resins and filler powders under pressure and heat; brass or zinc particles may again be included to improve the heat conductivity and modify the frictional properties.

Countersunk soft copper or aluminium rivets secure the linings to the brake shoes, which may be of pressed-steel, light-alloy or special cast-iron construction. Alternatively, the linings may be bonded on, usually using a thermosetting phenolic resin. This method allows an increased effective life for the lining and reduces drum scoring from foreign material accumulated in the rivet holes.

A ribbed construction in special alloy cast iron is usual for the brake drums, since it will withstand the brake-shoe pressures without distortion and has a low rate of wear.

Owing to the heat insulation of the friction material, 95 per cent of the heat generated passes into the drum; the remainder is confined to a surface layer of the linings. Under normal conditions this heat is satisfactorily dissipated but, after repeated stops from high speed or a prolonged descent with a heavy load, the contacting surface of the material can overheat and deteriorate. This can cause a marked fall in the coefficient of friction, which is normally 0.3–0.4 according to the material. While this brake fade can be largely overcome by the use of antifade linings, the associated drum expansion, bell mouthing, ovality and heat spotting upset the brake geometry and have led to the adoption of the disc brake on many vehicles.

22.4 Brake efficiency

Brake efficiency is generally expressed by giving the deceleration produced as a percentage or decimal fraction of g, the acceleration of a freely falling body (approximately 9.81 m/s^2).

The force acting on a freely falling body is its weight, and to produce an equal deceleration – that is, 100 per cent brake efficiency – requires a retarding force equal to the vehicle's weight. To obtain such a retarding force by the brakes necessitates a coefficient of friction μ not less than 1.0 between the tyres and the road, since $\mu = F/W$ and the retarding force F must equal the weight of the vehicle for 100 per cent efficiency.

In practice about 85 per cent is the safe maximum limit for cars, although figures over 100 per cent have been recorded on dry tarmacadam and imply an interlocking action between the tyre tread and the road. On public-service and heavy goods vehicles some 65 per cent is the safe maximum limit to avoid injury to passengers or damage to goods. The minimum requirements for the Department of Transport testing in the UK for cars is 50 per cent for footbrakes and 25 per cent for handbrake.

Brake efficiency may be assessed by measuring stopping distance: approximately 10 m from 50 km/h represents 100 per cent or $1.0g$ efficiency. From a given speed the stopping distance and stopping time is inversely proportional to the brake efficiency (e.g. 20 m from 50 km/h at $0.5g$). For a given efficiency the stopping time is directly proportional to the speed but the stopping distance is proportional to the square of the speed (e.g. 40 m from 100 km/h at $1.0g$ and 80 m from 100 km/h at $0.5g$).

In practice, the stopping distance depends upon the rapidity of the driver's reactions as well as brake efficiency; the driver's delay may vary from about one second with an unexpected stop to about half a second when prepared.

Adverse road conditions – ice etc. – may, of course, reduce the efficiency of normally effective brakes to some 10 per cent or less.

22.5 Brake actuation

The brakes can be operated mechanically by cables or rods, or hydraulically, or by a combination of these methods.

A considerable leverage is necessary in the actuating mechanism to give large forces at the brake shoes from a reasonable effort by the driver. A very large leverage, however, would necessitate an excessive pedal travel or constant adjustment to take up the lost motion of the brake-shoe clearance plus any stretch or wear in the linkage. The actual leverage must therefore be a compromise between these opposing conditions, and is of the order of 100:1.

22.6 Compensation

If the retarding forces on the nearside and offside wheels are unequal, there will be a tendency for the vehicle to swerve when decelerating. To avoid this, a balance lever or similar compensating arrangement is usually fitted in mechanical linkage to provide equal braking forces to the opposing wheels irrespective of any difference in the individual adjustment (Figure 22.3). Hydraulic operation gives inherent compensation since the fluid pressure is equal throughout the system.

Should the coefficients of friction for the braking surfaces be unequal – as, for example, when oil is present in one drum – then an unequal retarding force will be produced despite compensation.

22.7 Brake-shoe operation

Fixed-cam assembly

With the simple fixed-cam-operated assembly, the force of friction produces a turning moment about the brake-shoe anchor pins which tends to rotate the leading shoe towards the brake drum and the trailing shoe away from the drum. As a result, the cam exerts a greater force on the trailing shoe than on the leading shoe (Figure 22.4).

Floating-cam assembly

By permitting the expanding mechanism to float slightly on the back plate, an equal force is applied to each shoe and a greater braking torque is produced for a given actuating force. With this system, however, the lining wear on the leading shoe is greater than on the trailing shoe and sometimes a thicker lining is fitted (Figure 22.5).

Two-leading-shoe assembly

By using the two-leading-shoe (2LS) arrangement, an increased braking torque for the same actuating force is obtained, and the lining wear is equal since the force of friction tends to rotate both shoes about the anchor pins towards the brake drum. The effects of brake fade are,

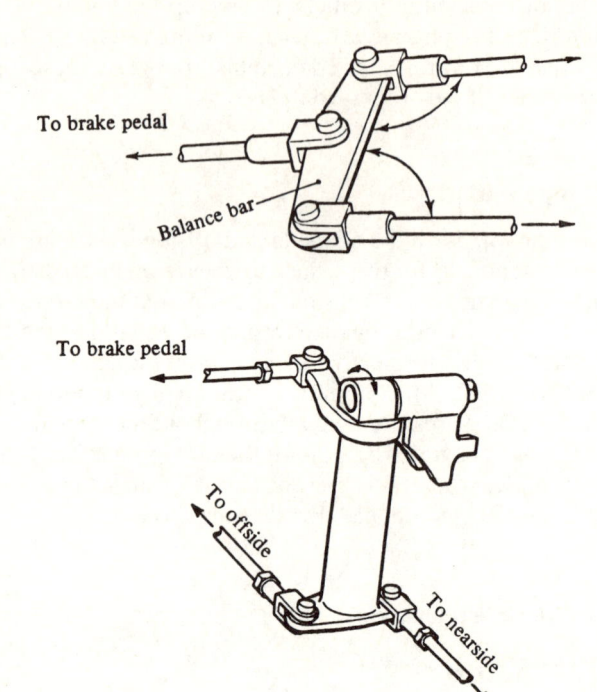

Figure 22.3   Brake compensators

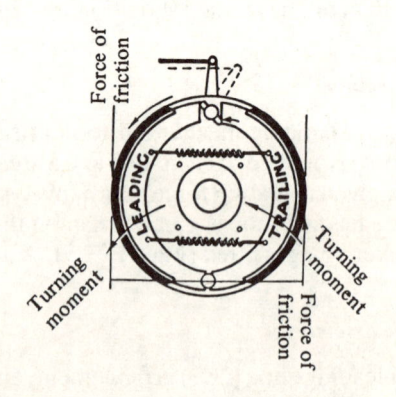

Figure 22.4   Fixed-cam assembly

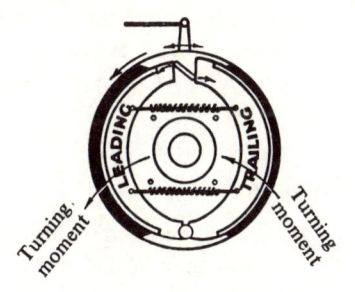

Figure 22.5 Floating-cam assembly

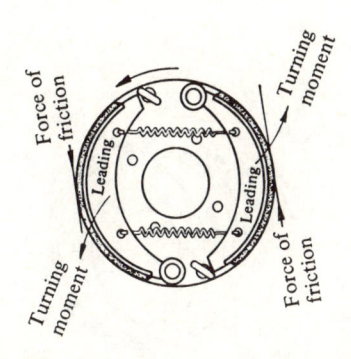

Figure 22.6 Two-leading-shoe assembly

however, correspondingly increased by the self-wrapping action of the two-leading-shoe system (Figure 22.6).

In some cases, the anchor ends of the shoes are not pivoted but have inclined flat abutments. The combination of the expander thrust and the force of friction causes the shoes to slide outward at the anchor end, equalizing the pressure, heat and wear over the whole lining surface and increasing the braking torque.

Trailing shoes

The two-leading-shoe arrangement becomes a two-trailing-shoe brake if the direction of the drum rotation is reversed, and this necessitates an increased actuating force for a corresponding braking torque, normally provided by a vacuum servo unit. The trailing-shoe system has the advantage, however, that the braking torque is considerably less affected by a reduction in the coefficient of friction of the linings, since it does not rely upon the turning moment produced by the force of friction on the linings. Brake fade is therefore less pronounced with the trailing-shoe brake.

Floating-anchor, fixed-cam assembly

If the leading and trailing shoes are linked together and allowed to float on the back plate at the usual anchor-pin ends, the force of friction acting on the leading (or primary) shoe and tending to rotate it with the brake drum applies an increasing braking force on to the trailing (or secondary) shoe (Figure 22.7).

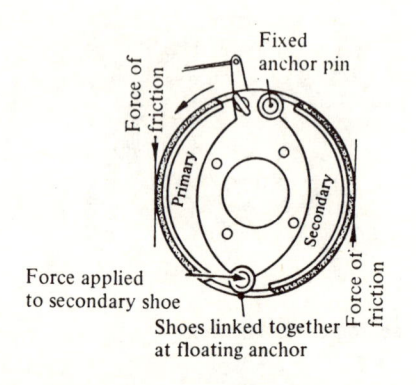

Figure 22.7 Floating-anchor assembly

By this self-energizing action the braking torque is considerably increased, but it is sensitive to any variation in the coefficient of friction and care is needed in the selection and fitting of the linings.

The braking torque for the different shoe assemblies is as follows:

Shoe operation	Approximate relative braking torque
Fixed expander and anchor	1.0
Floating expander, fixed anchor	1.5
Two leading shoes, fixed anchors	2.0
Fixed expander, floating anchor	3.5
Trailing shoes, fixed anchors	0.5

22.8 Brake torque

The back plate carrying the brake shoes tends to rotate with the brake drum when the brakes are applied, and this torque must be transmitted to the vehicle structure through leaf suspension springs, a torque tube or by the independent-suspension linkage.

22.9 Mechanical operation

Linkage

Mechanical linkage on cars is confined to the handbrake action, where safety and legislation require an independent operation of the brakes. Some means of compensation between the offside and nearside wheels is needed – for example, a pulley-and-cable, balance-lever or swing-link connection.

Expander

Mechanical expanders are required for the handbrake. With hydraulic actuation, they are also fitted to light commercial vehicles. By distancing the cylinder and piston from the friction surfaces, they allow a reduction in brake-fluid temperature. Heavy vehicles use a compressed-air cylinder to actuate mechanical brake-shoe expanders.

An alternative to some form of cam is to use a wedge to force apart two plungers acting on the brake shoes (Figure 22.8). Rollers interposed between the wedge and the plungers minimize friction. The wedge angle

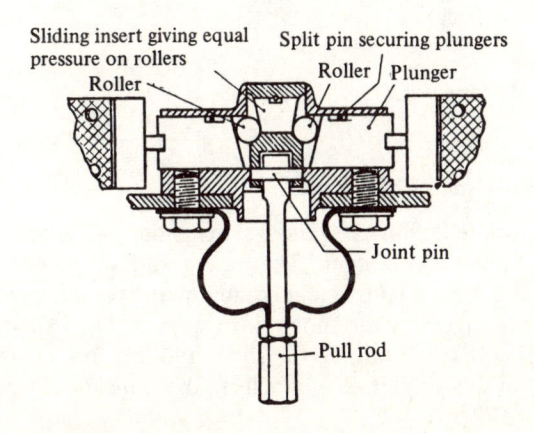

Figure 22.8 Wedge expander

usually gives a ratio of about 6:1 between wedge and shoe movement. In some designs the wedge can float to equalize the brake-shoe forces. The principle can be combined with hydraulic operation for the rear brakes of light commercial vehicles. The wedge is pulled hydraulically for the footbrakes and mechanically for the handbrake. For the front brakes the wedge can be reversed and pushed by a direct-acting hydraulic cylinder. In both applications, mounting the hydraulic cylinders outside the brake drums lowers the fluid temperature.

Wedge adjuster

The adjuster unit, bolted rigidly to the back plate, contains two plungers on which the brake shoes pivot and which can be forced apart by a screwed conical wedge (Figure 22.9). Flats on the operating surface of the adjuster allow the brake-shoe return springs to lock the adjustment and also provide an accurate brake-shoe clearance. When adjusting, the wedge is screwed in until the shoes contact the drum, and then slackened back to allow the plungers to seat on the flats on the cone face and so give a predetermined clearance.

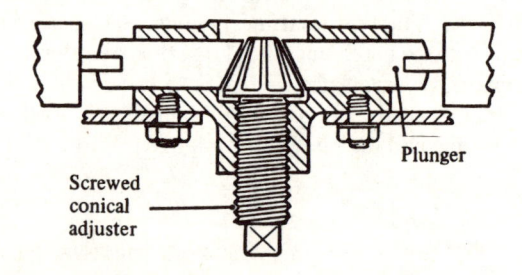

Figure 22.9 Wedge adjuster

Mechanical two-leading-shoe brake

Twin-leading-shoe action using a single expander can be achieved using a strut-and-bell-crank system. The brake shoes are not positively located but are held against inclined abutments on the expander unit and adjuster plungers by the shoe return springs. The expander tappet acts on a bellcrank pivoting on the shoe, and this force is transmitted through a strut to a similar second bell crank acting against the adjuster tappet (Figure 22.10).

With the drum stationary, both ends of the shoe are slightly lifted

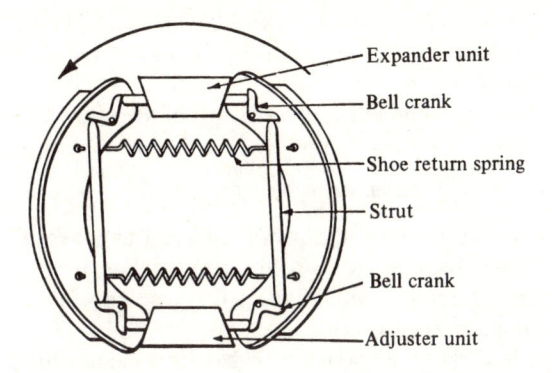

Figure 22.10 Mechanical two-leading-shoe brake

from their abutments; rotation causes the shoe to move against one abutment and become a leading shoe in either direction.

For normal two-leading-shoe action only one strut-and-bell-crank system is needed, fitted to the trailing shoe; fitted to both shoes the system becomes a double-acting two-leading-shoe type, effective in both directions.

22.10 Hydraulic operation

The brake pedal operates a piston, forcing fluid from the master cylinder into the pipeline system; this is already completely filled with fluid, and an increase in pressure is therefore produced. This pressure acts on pistons in the wheel cylinders which are in contact with the brake shoes, and so forces the shoes against the brake drums. The fluid operation eliminates the friction of mechanical linkage and gives inherent compensation. In addition, wheel movement is readily accommodated by flexible hoses.

The force on the brake shoes can be increased by using larger-diameter wheel cylinders and pistons.

Hydraulic operation has certain disadvantages:

(a) Air, unlike a liquid, is readily compressible, and if it is allowed to enter the system it will reduce the efficiency and give a spongy pedal action.
(b) A serious leakage in any part of the system will cause a complete hydraulic failure unless a protective design is adopted.
(c) A separate connection from the handbrake to the brake shoes is necessary.

(d) The design of the wheel cylinders must ensure that acceptable brake-fluid temperatures are not exceeded during prolonged braking.
(e) Periodical topping up and exchange of the brake fluid is required.

Master cylinder: residual-pressure type

In the returned position the piston cup seal uncovers a port, allowing fluid to enter from or return to the reservoir as its volume varies with expansion. The brake pedal has a small free movement to ensure that the piston uncovers this port.

Delivery to the pipeline is through an inner non-return valve (Figure 22.11). When the brake pedal is released and the fluid is returned to the master cylinder by the action of the brake-shoe return springs, it must raise the valve assembly against the tension of the piston return spring, and this maintains a pressure of about 55 kPa in the pipeline. This residual pressure ensures that cup seals in the wheel cylinders are expanded and exclude air.

An annular space behind the piston head is supplied with fluid through a separate drilling from the reservoir. On the return stroke of

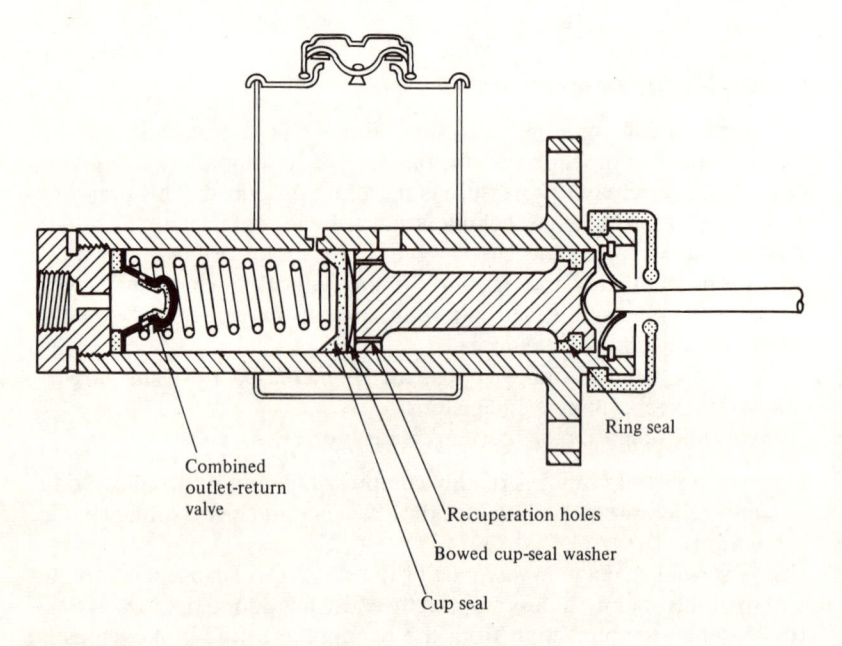

Figure 22.11   Master cylinder: residual pressure

the piston, the reduced pressure in the pumping chamber allows fluid to pass through the holes in the piston head, around the cup seal and into the chamber. This recuperative action ensures sealing and enables additional fluid to be rapidly transferred into the pipeline by pumping the brake pedal.

The reservoir, which may be integral with the master cylinder or separately mounted, has a valve in the filler cap to allow air to enter if the pressure falls below atmospheric. A fluid-level warning-light sensor float may be fitted to the cap.

For some purposes – disc brakes and self-adjusting hydraulic clutch operation – the system must not have any residual pressure in the pipelines, and the combined outlet-and-return check valve is omitted or replaced by a modified type with a small bleed hole.

Master cylinder: non-residual-pressure type

In the returned position of the plunger the valve seal is retracted within the valve spacer by the action of the spring retainer and valve shank, allowing a free flow of fluid between the pipeline and the fluid reservoir.

The initial movement of the plunger releases the valve shank and, with the valve spacer in contact with the end face of the cylinder, the seal is forced on to its seat by the spring washer (Figure 22.12). Continued movement displaces fluid into the pipeline, the valve shank passing further into the hollow centre of the plunger.

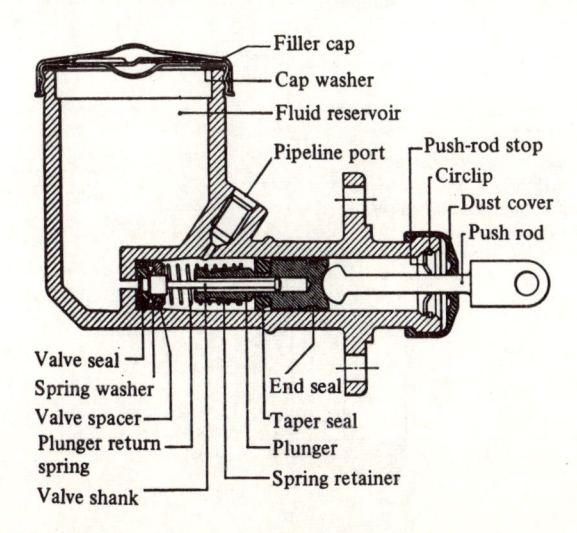

Figure 22.12 Master cylinder: non-residual pressure

Wheel cylinder

The original type, open at each end, contains two pistons acting on the shoes, giving an equal force on each and therefore greater lining wear on the leading shoe. Two cup seals with a light spring between them are used to provide sealing, and rubber boots keep the cylinder free from dust (Figure 22.13).

The two-leading-shoe system employs a separate cylinder (closed at one end) and piston for each shoe, the two cylinders being linked by a fluid pipe (Figure 22.14).

For the rear brakes a single-acting cylinder may be used; the piston bears on the leading shoe, whilst the body of the cylinder, permitted to float on the back plate, acts on the trailing shoe (Figure 22.15). Mechanical operation through the handbrake linkage is by a bell-crank lever bearing on the outer portion of a divided piston, or directly on to

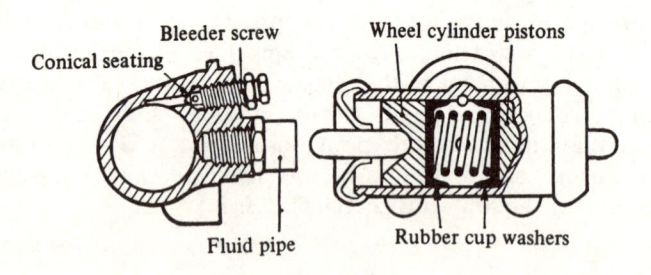

Figure 22.13   Two-piston wheel cylinder

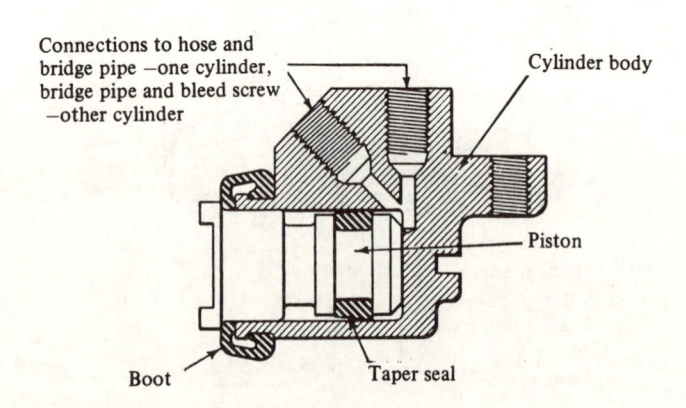

Figure 22.14   Single-piston cylinder for two-leading-shoe brake

The braking system

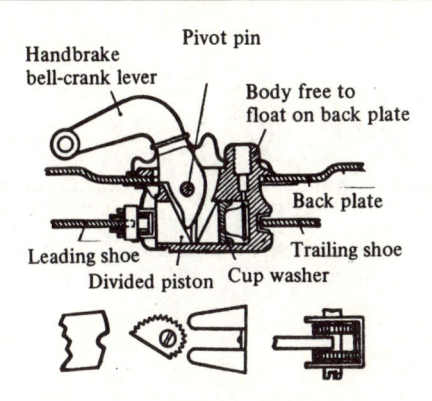

Figure 22.15 Single-acting cylinder and adjuster

the brake-shoe web. In each case, the trailing shoe is operated by the reaction of the pivot pin carried in the cylinder body. Alternatively, a double-piston wheel cylinder may be used, with handbrake operation through a bell-crank and strut, or a wedge expander acting directly on to both shoes, adjustment being at the anchor end by a screwed wedge (Figure 22.16).

To avoid excessive expansion or vaporization of the fluid, the operating cylinder can be mounted outside the brake drum (Figure 22.17). A wedge-type expander can be used, operated by the hydraulic piston for the foot brakes and by rod for the handbrake.

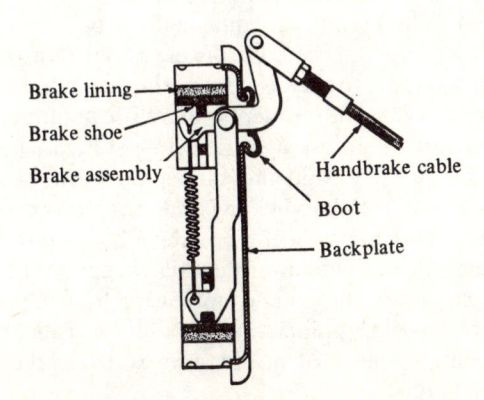

Figure 22.16 Handbrake: bell crank and strut

283

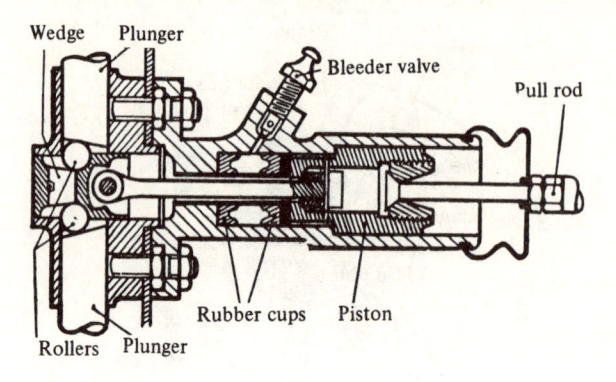

Figure 22.17   External wheel cylinder

22.11   The disc brake

The wheel hub carries a disc of special cast iron, which is either flat or of top-hat section and which may have internal cooling vents. The caliper, secured to the stub-axle or rear-hub assembly, straddles the disc (Figure 22.18). In earlier systems, a coaxial cylinder on each side contains a hollow piston with sealing ring and dust excluder. Inserted between the pistons and the disc are the friction pads, bonded to steel backing plates, which abut in recesses in the caliper.

When the brakes are operated, fluid from the master cylinder passes to both pistons and creates an equal and opposite force on each pad. When the brakes are released, the pistons and pads are retained by the sealing rings adjacent to the disc; retraction is limited to the flexure of the sealing rings. Wear adjustment is therefore automatic.

Current practice is to use a single-acting system where the inboard piston acts on one pad and the caliper body on the other. This is economical and avoids fluid heating and location problems within the shrouded dish of the road wheel. The caliper body must pivot or slide on its anchorage. Pivotal mounting entails wedge-shaped wear as the caliper swings; this is sometimes counteracted by using pads of an initial wedge section. Sliding calipers must avoid the risks of corrosion and seizure preventing release of the pressure exerted by the caliper body on the outboard pad.

The master cylinders used are of the non-residual-pressure type.

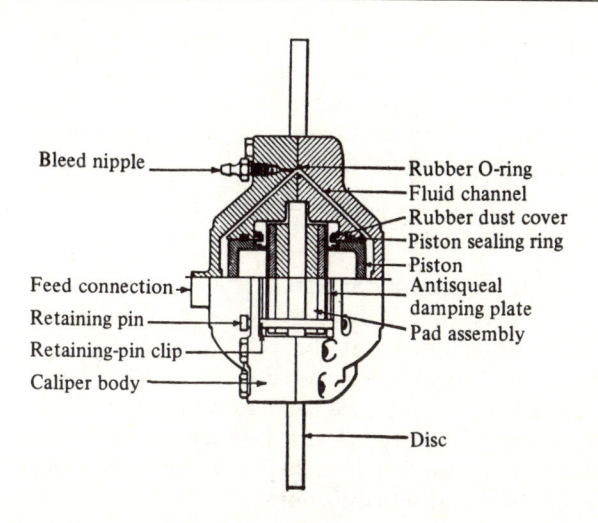

Figure 22.18   Twin-piston disc-brake caliper

Advantages of the disc brake

(a) The friction surfaces are exposed to the air stream except for the pad areas.
(b) Although high temperatures can be reached with discs, radial expansion and the negligible lateral expansion have no effect on brake geometry.
(c) Owing to manufacturing differences a greater range of ingredients can be used in friction pads than in linings, permitting a wider choice of characteristics.
(d) Lack of the self-wrapping action of the two-leading-shoe brake results in more consistent braking, less affected by fade and equally effective in reverse.
(e) Being self-adjusting with negligible clearance, a high mechanical advantage can be obtained in the hydraulic system.
(f) Easy inspection and renewal of the friction material.

Disadvantages of the disc brake

(a) Although larger piston areas are used than for drum brakes, the lack of a self-wrapping action usually necessitates a vacuum servo to maintain the usual limit of 225 N pedal force for $0.5g$ retardation.

(b) Handbrake operation presents difficulties.
(c) Mud, water and dirt are largely thrown off by centrifugal action and the wiping effect of the pads, but under adverse conditions excessive or unequal inboard and outboard pad wear, disc corrosion or scoring, and piston seizure can occur.

22.12 Brake layout

Typical layouts, taking advantage of static weight distribution and weight transfer, are disc or twin-leading-shoe brakes at the front and single-leading-shoe brakes at the rear (Figure 22.19). The latter, adequate for the limited braking force available, are equally effective in reverse and avoid a complicated handbrake application to a disc.

For light commercial vehicles, a twin-leading-shoe front assembly may be combined with double-acting twin-leading-shoe rear brakes to give effective action in both directions.

Front-wheel-drive cars utilize some device to prevent the lightly loaded rear wheels locking under weight transfer when braking. On some a pressure-reducing valve restricts the maximum pressure in the rear line to, say, 3.5 MPa, and additional pressure is applied to the front brakes only. Alternatively, a hydraulic valve can be linked to the suspension and will operate under brake dive. These simple devices are limited in refinement, although later developments incorporate rear-wheel loading. It seems that a comprehensive antilock system – where wheel deceleration is electronically sensed and appropriate signals relieve braking pressure in a rapid cycle – will prove most effective.

Inboard brake

Inboard location of the brakes with an independent-suspension system has the advantage of reducing the unsprung mass of the wheel assemblies, relieving the suspension mountings of braking torque, and simplifying the brake linkage from suspension and steering movement.

The drive shaft and its couplings must transmit the brake torque, which in speed of application and magnitude may frequently exceed the driving torque.

Divided systems

Divided systems, enforced by legislation in most countries, are necessary to prevent a single leakage causing total brake failure. A common method is to use a tandem master cylinder where a floating piston divides the pumping chamber into two sections (Figure 22.20). The

The braking system

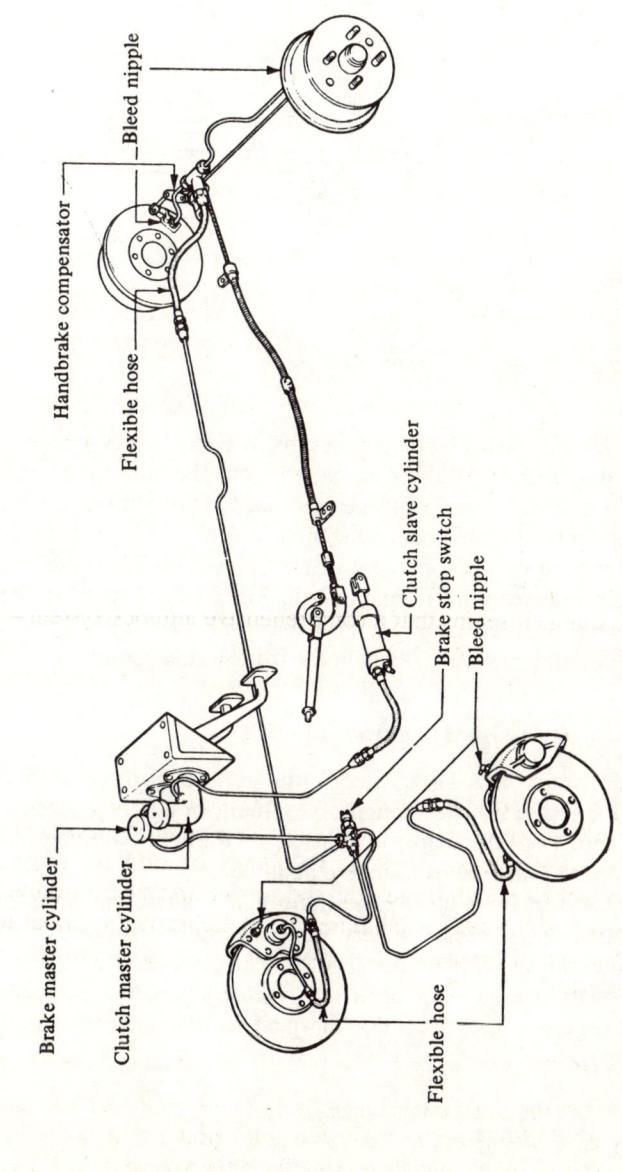

Figure 22.19 Disc and single-leading-shoe drum single-line hydraulic circuit

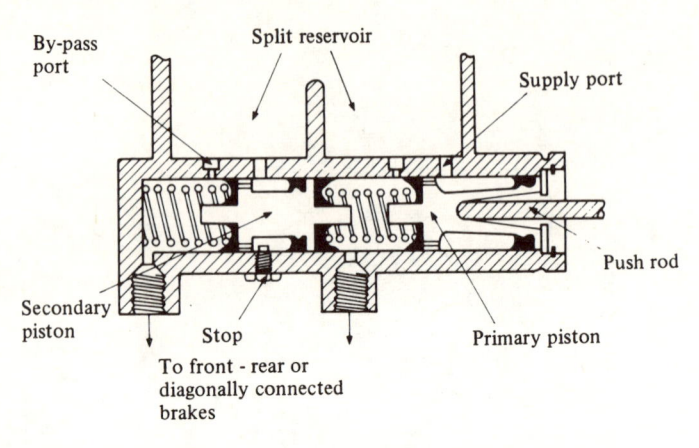

Figure 22.20   Tandem master cylinder

floating piston ensures equal pressure in both halves but, should a failure allow a loss of fluid from one section, the fluid in the other can still be pressurized and will operate that half of the system, though requiring an additional pedal movement.

Unless some of the components are to be duplicated, the choice of division is between a front–rear or a diagonal split. The front–rear split provides only limited efficiency from the rear brakes only. The diagonal division requires negative offset in the front-wheel geometry.

22.13   Servo-assisted braking

For a constant braking efficiency the retarding force increases in proportion to the vehicle's weight. Eventually brake forces are required which cannot be obtained from the driver's effort; 400 N is normally considered the maximum sustained pedal force.

For cars with front disc brakes, and light commercial vehicles with front drum brakes, a servo mechanism is generally employed to enhance the driver's effort. Should the servo fail, the brakes can be less effectively applied by unaided pedal action. Heavier vehicles, especially with trailers, require power operation in which the driver exerts a controlling action only.

Vacuum servo systems

Vacuum servo operates from the depression in the induction manifold when a spark-ignition engine is running – say 2 kPa depression.

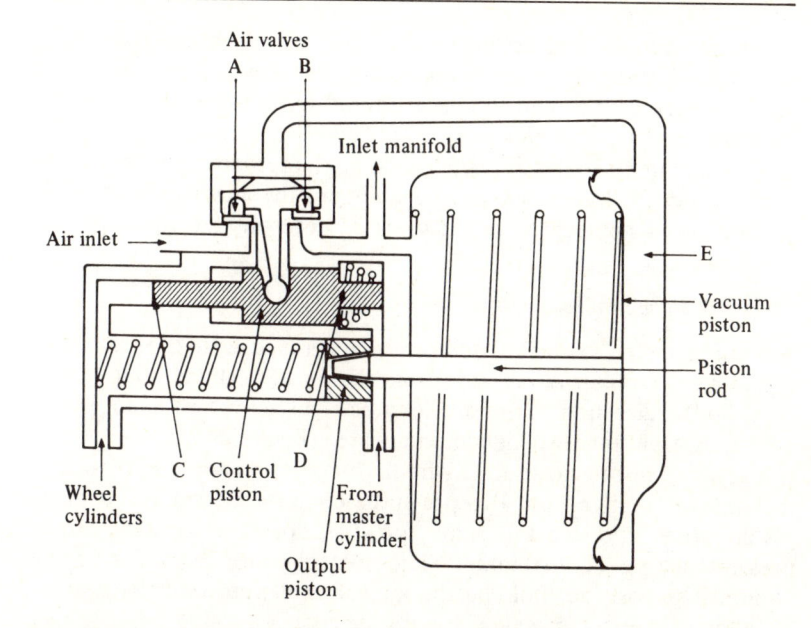

Figure 22.21   Suspended-vacuum indirect-type servo
(a) Brakes off. Valve A closed, B open. Suspended-vacuum condition.
(b) Brake pedal operated creating an equal hydraulic pressure at C and D. Larger area of D causes control piston to move left, opening A, closing B. Air enters E, vacuum piston moves left, piston rod seals output piston and augments hydraulic pressure to wheel lines. When augmented hydraulic pressure at C acting on area C has the same force as the master-cylinder pressure acting on area D the control piston centralizes, closing valves A and B. The brakes are held on with a pressure increase in the ratio area D/area C.
(c) Any increase or decrease in master-cylinder pressure is reproduced proportionately in the wheel line pressure.
(d) Brake pedal released. Control piston moves right, closing valve A and opening B to restore condition 1.

Commercial vehicles with compression-ignition engines use an exhauster pump, check valve and vacuum reservoir; CI cars normally have a small diaphragm pump operated from an eccentric on the camshaft.

On earlier servo systems both sides of a vacuum piston were at atmospheric pressure when at rest. In current applications the depression acts on both sides of the piston or diaphragm until brake operation admits atmospheric air to one side – at 100 kPa. This suspended vacuum gives a prompt response and avoids a large volume of air suddenly entering the inlet manifold – which occurred with the older system when the brakes were applied.

A reaction or balance valve, operated mechanically or by the pressures in the hydraulic system, maintains the augmented thrust in proportion to the pedal force and gives the necessary feel to the pedal.

Suspended-vacuum indirect-type servo

A floating control piston has differential areas exposed to the brake fluid in the rest position. A pressure from the master cylinder causes it to move in the direction of the smaller area. This activates the air valves, admitting air to the suspended vacuum area (Figure 22.21).

The servo piston closes the hydraulic by-pass and applies pressure to the pipeline system until this pressure acting on the smaller control-piston area produces the same force as does the master-cylinder pressure acting on the larger. The control piston balances in this equilibrium position, holding the air valves closed until the master-cylinder pressure changes.

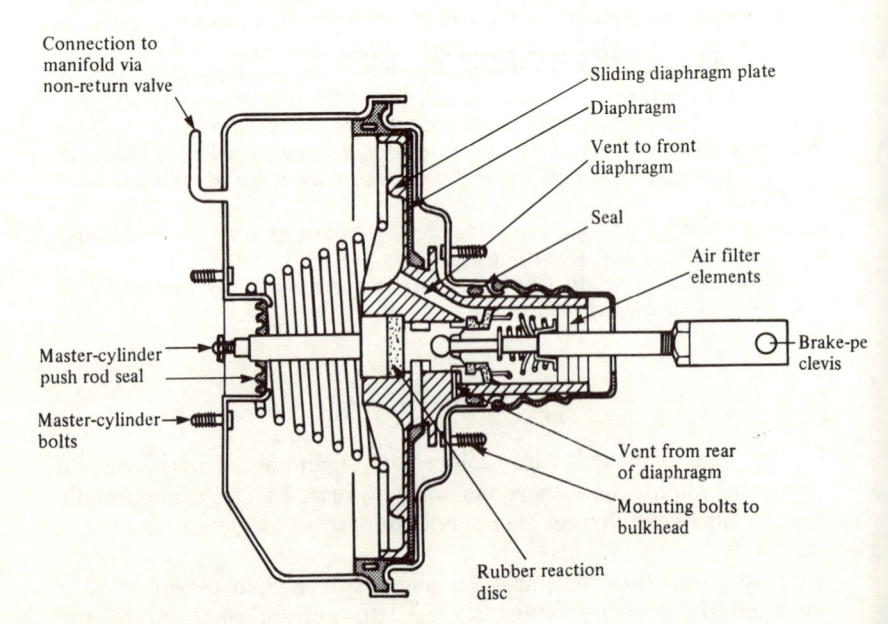

Figure 22.22 Suspended-vacuum direct-type servo

The braking system

Suspended-vacuum direct-type servo

In this type of direct-acting servo unit, the valve control system is housed in the moving diaphragm plate. The general principle is similar to the previous type but the method of apportioning servo assistance to pedal force is by a rubber reaction disc. This disc is trapped between the diaphragm plate and the master-cylinder push rod (Figure 22.22).

Servo pressure causes the diaphragm to deform and bulge against the head of the air valve operated from the brake pedal (Figure 22.23). When these two forces are in equilibrium the air valves remain closed

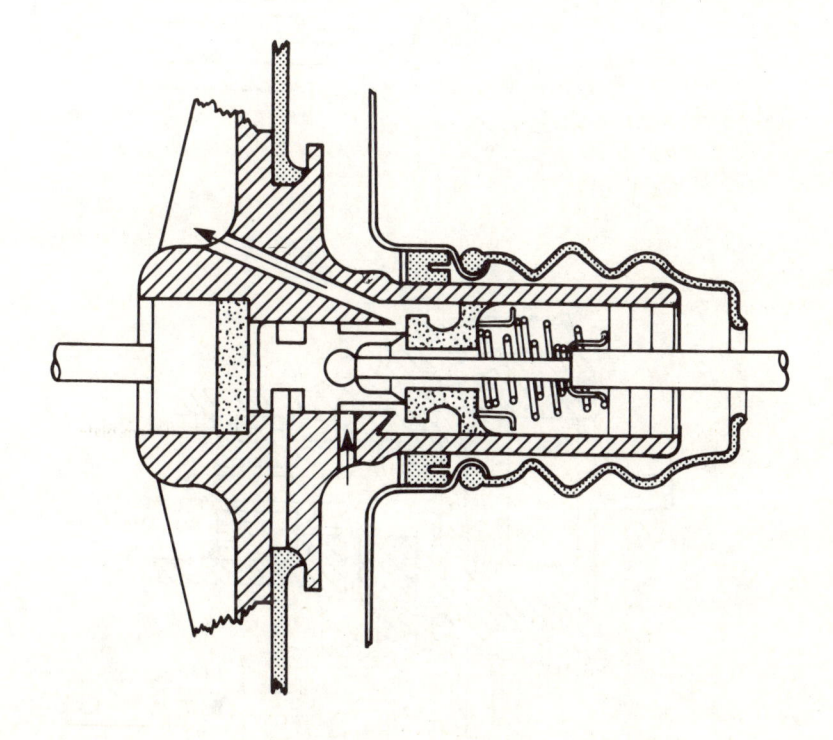

Figure 22.23 Valve action on direct-type servo
(a) Brakes off. Air exhausted from both sides of diaphragm plate. Suspended-vacuum condition.
(b) Brake pedal operated. Air admitted, causing servo diaphragm and valve assembly to move to the left, applying force through rubber reaction disc to master-cylinder push rod.
(c) Equilibrium condition. Force from deformed reaction disc balanced by pedal force. Air valves closed. Situation (a) or (b) follows depending upon pedal force decreasing or increasing

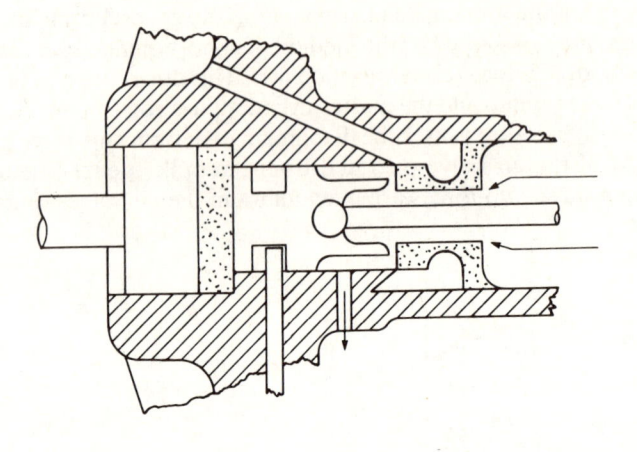

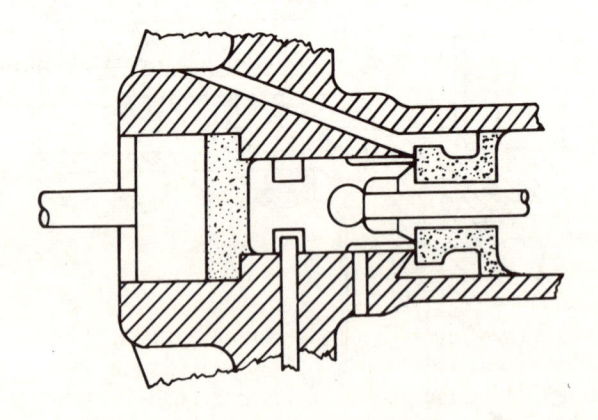

until the brake-pedal force changes. The force on the master-cylinder push rod is thus approximately proportioned to the brake-pedal force.

22.14 Compressed-air power braking

In a simplified arrangement, a single- or twin-cylinder compressor, driven from the engine or transmission, charges a compressor through a check valve. When the working pressure (some 700–800 kPa) is reached, a governor valve activates an unloader valve on the compressor, lifting the inlet air valves and allowing it to run light. A filter and antifreezer are fitted on the compressor air intake; the latter, by feeding small quantities of methanol, lowers the freezing point of any moisture in the air.

The reservoir is divided into two compartments by a diverter valve which allows the pressure to be rapidly built up in the smaller section. The larger capacity of the main chamber ensures a reservoir for several brake applications.

The brake treadle controls a reaction valve in the service pipeline to the wheel actuators, thus apportioning brake operating pressure to the treadle force. The actuators – usually diaphragm rather than piston type – operate a mechanical system for brake-shoe application.

Some systems use a safety arrangement where powerful springs are held in compression by air pistons. Should the air pressure to these pistons be released or fail, the springs actuate the brakes.

Handbrake action is through a separate air system, but once the brakes are applied they are automatically held by the mechanical action of springs (as above) or by a mechanical lock, released by air pressure when required.

A trailer carries an auxiliary reservoir and relay valve whereby the trailer brakes are automatically applied should the air line become detached.

22.15 Auxiliary braking

Auxiliary brakes – retarders – are fitted to heavy vehicles to provide a reasonable degree of retardation without using the main brakes, which are available for additional or emergency use. They are particularly useful on long declines where the gradual accumulation of heat can cause brake fade. In addition they reduce tyre wear, risk of skidding, fuel consumption in the case of exhaust braking, and driver fatigue.

The exhaust brake employs a valve operated by solenoid, vacuum or compressed air in a chamber near the exhaust manifold. This restricts the gas flow and builds up a pressure, limited to some 300 kPa by the valve setting, whilst at the same time fuel injection is cut off. The engine absorbs the kinetic energy by acting as a low-pressure air compressor.

A typical eddy-current brake, located in the propeller-shaft line, comprises two air-cooled discs revolving each side of chassis-mounted electromagnets. As the rotors cut this magnetic field, eddy currents are induced in them and oppose their rotation. The control lever determines the number of coils energized from the vehicle electrical system and hence the retardation.

22.16 Maintenance of brakes

Adjustment

Self-adjustment, with the resulting small operating movements of the shoes, enables a high leverage to be employed in the actuating system with reduced pedal forces. On drum brakes the system is complicated by drum expansion and the self-wrapping action of the leading shoes. Car systems may employ a simple frictional retention of the shoes near the drum for the rear brakes. Alternatively, a screwed adjuster can be used, which is advanced by a ratchet actuator, whenever the brake shoe movement exceeds the designed maximum.

Snail cams, screwed tappets and screwed wedges provide for manual adjustment. In some cases provision is also made to centralize the shoes in the drum.

After prolonged use, cable or rod handbrake linkage may need resetting to obtain the original leverage. Maximum torque on a brake operating lever is obtained when the rod or cable makes a right angle with the effective arm.

Linings and pads

The resistance of friction materials to fade, wear, oil or water contamination varies considerably. The coefficients of friction for the materials may range from under 0.3 to over 0.4, whilst limiting temperatures can be below 300°C or above 700°C. It is therefore necessary for efficiency and safety that the correct material is used when replacing pads or linings. Pads should be inspected for unequal wear every 10 000 km; drums should be removed to examine the linings every 20 000 km. Note that brake dust is a health hazard.

Replacement is indicated when the lining approaches the rivets, or when the material thickness on bonded linings or pads is approximately 1.5 mm. Some pads incorporate a warning-lamp sensor wire at the depth of permissible wear.

Lubrication

Special greases – for example, containing zinc oxide – are available for brake-gear lubrication, which avoid drying out and corrosion. These must not contact rubber components; special rubber lubricants are available for these areas.

Hydraulic system

Maintenance of the fluid level is necessary to prevent air entering the system. The reservoir level should be checked every 5000 km and the cause of any substantial loss of fluid established. A small transfer of fluid must occur as the caliper pistons take up disc-pad wear. A commonly used glycol-base fluid is hygroscopic and, since water absorption lowers the boiling point, the manufacturers recommend changing the fluid at intervals of two to three years.

Should air have entered the system, bleeding will be necessary. The basic procedure is to unscrew the bleeder nipple about half a turn and operate the brake pedal with slow full strokes, allowing it to return freely. The displaced fluid passes from the nipple through an attached pipe into a clean jar and the operation is repeated until the flow is free from air bubbles. The nipple is then tightened – to a torque of some 5–10 Nm – on a downward stroke of the pedal. The operation is repeated at the other bleed points, care being taken to keep the reservoir recharged with new fluid.

With dual-circuit systems this basic operation has often to be modified by varying the pedal action or by using a pressurized fluid connection to the reservoir and a prescribed sequence of bleeding.

22.17 Brake defects

Brake grab or judder

(a) Worn or contaminated pads or linings
(b) Loose mountings, i.e. anchor pin, back plate, caliper, suspension; steering or wheel bearings loose
(c) Scored, distorted or corroded discs or drums
(d) Pistons in caliper or wheel cylinder seized or tight
(e) Pads or shoe linkage seized or tight
(f) Incorrect type of friction material
(g) Defective servo unit.

Brake pulling to one side

(a) Judder defects affecting one disc or drum
(b) Hydraulic compensation defective, i.e. blockage, divided system defective
(c) Unbalanced tyre pressures.

Brake drag

(a) Adjustment too tight, automatic adjuster faulty, handbrake linkage seized or incorrectly adjusted
(b) Pistons or pad assembly seized
(c) No free movement of brake pedal – piston prevented from returning fully
(d) Faulty servo action
(e) Hose or pipe blocked, seals damaged by contamination, reservoir overfilled, vent blocked
(f) Loose wheel bearings
(g) Incorrect master-cylinder check valve creating residual pressure in disc brake system.

Localize defect by slackening the bleed screw; if the wheel still drags the fault lies in the brake assembly.

Excessive pedal travel

(a) Brake-shoe adjustment required, faulty automatic adjustment
(b) Knock-back of pads and pistons by excessive disc run-out – normally 0.1 mm maximum – or loose wheel bearings, indicated by pedal flutter when applied
(c) Distorted damping shim forcing back caliper piston
(d) Fluid leakage.

Spongy pedal action

(a) Air trapped in hydraulic system
(b) Fluid vaporization.

High pedal force needed

(a) Faulty servo
(b) Seized caliper or wheel cylinder pistons
(c) Worn, contaminated, glazed or incorrect type of friction material.

Brake squeal

(a) Pads or linings new and not bedded, contaminated, dust in drums
(b) Damping shims defective
(c) Loose caliper assembly, back plate
(d) Scored or distorted disc or drum
(e) Loose wheel bearings.

Example 24

A car has a wheelbase of 2.134 m (Figure 22.24). The rear wheels have a static load of 7.620 kN and the coefficient of friction between the tyres and the road is 0.65. If the centre of mass is 0.460 m above the ground, calculate the increase in load on the front axle when the rear brakes are applied to their maximum capacity without skidding.

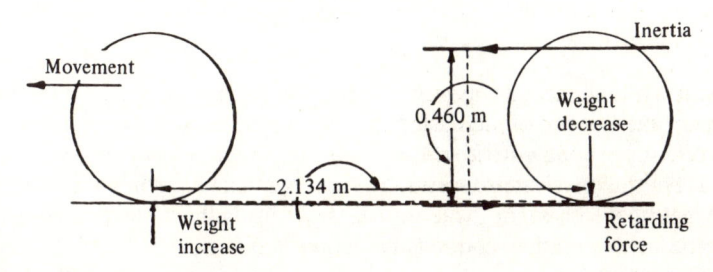

Figure 22.24

The static load on the rear axle = 7620 N. Let the transferred weight under maximum deceleration be W N. The load on the rear axle under these conditions = $(7620 - W)$ N.

Now

maximum retarding force on rear wheels
= force of friction between tyres and ground
= $\mu \times$ normal load on axle
= $0.65 (7620 - W)$ N

Therefore

overturning couple formed by retarding force and inertia of vehicle
= $0.65 (7620 - W) \times 0.460$ Nm
balancing weight transfer couple = $W \times 2.134$ Nm

Hence

$$2.134 \times W = 0.299 (7620 - W)$$
$$= 2278.38 - 0.299W$$
$$2.433 \times W = 2278.38$$
$$W = 936.4 \text{ N}$$

There is an increase in the normal force on the front axle of 936 N under these circumstances.

Chapter 23
Materials

For most engineering purposes, pure metals are too weak. The most important method of increasing their strength, toughness or hardness is by blending them with other metals or non-metals to form alloys.

Often the physical properties of the alloys can be further improved by correct heat treatment, which consists of raising the metal to a required temperature and then cooling at a definite rate.

Ferrous metals are those which contain a large proportion of iron (Latin *ferrum*), and their varied physical properties depend on the effects of carbon and certain other elements alloyed with the iron.

23.1 Steel

Steel is basically an alloy of iron and not more than 1.5 per cent carbon, which is all chemically combined with the iron. When a greater proportion of carbon is present it cannot all be chemically combined and some will be in the form of free graphite; the metal is then a cast iron.

Table 23.1 shows the changing colours of steel with temperature.

Table 23.1

Colours of steel	*Temperature* (°C)
Very dull red	500–600
Dark blood-red	600–700
Cherry-red	700–800
Bright red	800–850
Pinkish-red	850–900
Pinkish-orange	900–950
Orange	950–1000
Yellow	1000–1100
Yellow-white	1100–1200
White	1200–1300
Brilliant sparkling white	over 1300

23.2 Mild steel

Steel containing up to about 0.25 per cent carbon is termed mild steel. It is very widely used where ductility with reasonable strength and toughness are important, for example in bodywork, tubing and brackets. Tinplate or 'tin' consists of mild-steel sheets with an anticorrosive thin coating of tin. Galvanized iron is also mild steel, with a protective coating of zinc.

23.3 Carbon steels

Medium-carbon steel contains from 0.25 to 0.5 per cent carbon; with 0.5 to 1.5 per cent it is usually termed high-carbon steel. Tool or cast steel is a high-carbon steel with 1 per cent or more carbon; silver steel is about a 1 per cent carbon steel accurately finished to size and polished.

The principal characteristic of the carbon steels is their property of hardening when quenched from a red heat. The hardness and strength increase, but with a loss in ductility, as the proportion of carbon rises to about 1 per cent. From 1 to 1.5 per cent of carbon there is a further increase in the hardness but some reduction in the strength of the steel.

23.4 Heat treatment of carbon steel

When carbon steel is heated to a temperature of 780–840°C (depending upon the composition) – i.e. above the 'critical' temperature – and then quenched it becomes very hard and brittle. For most purposes this brittleness must be reduced and the metal toughened by tempering.

Tempering consists of reheating the metal to a much lower temperature, determined by the intended use of the steel. In the workshop this temperature is often judged by the colour of the oxide formed on the surface of the heated metal. For example, a turning tool may be tempered to a straw colour – about 225°C – whilst a spring would need to be raised to a dark blue – about 300°C – to remove sufficient brittleness and give flexibility. Table 23.2 shows the carbon-steel tempering colours.

Annealing is carried out by heating slowly to a red heat (the annealing temperature varies slightly with the carbon content of the steel), 'soaking' or maintaining at this temperature for a time dependent on the size and shape of the article, and then cooling as slowly as possible – either inside the furnace or else buried in sand or ashes. This treatment renders the steel soft and more easily machined and also relieves internal stresses which may have been created by cold working or unequal contraction when cooling.

Table 23.2 Carbon-steel tempering colours (of the oxide)

Appearance	Temp. (°C)	Purpose
Light straw	220	Scrapers, brass-turning tools
Straw	225	Turning and parting tools
Dark straw	230	Hammer faces
Yellow-brown	245	Dies, reamers, boring and milling cutters
Brown	250	Taps, pen-knives, chasers
Brown-purple	265	Twist drills
Purple	275	Press tools, axes
Violet	290	Cold chisels
Blue	295	Screwdrivers
Dark blue	300	Springs

Normalizing consists of cooling from a red heat in air; the rate of cooling is therefore intermediate between quenching, where the object is cooled very rapidly, and annealing, where it is cooled as slowly as possible. Normalizing relieves internal stresses in the steel and refines its structure, which may have been coarsened by prolonged heating, such as is required for some forging operations.

23.5 Case hardening

Quenching from a red heat has no appreciable effect on mild steel owing to the low carbon content. Steel has the property of absorbing carbon when raised to about 900–950°C, and if mild-steel articles are packed in boxes containing charcoal or other substances rich in carbon, and maintained at this temperature for several hours, the carbon content of the surface layers of the metal is considerably increased. This high-carbon case can be hardened by quenching in water and, if required, tempered. The resulting article combines a tough mild-steel core with a wear-resisting hardened skin.

In the workshop a quick method of giving a superficial case on small mild-steel articles is to heat them to redness and plunge into a suitable hardening powder; the items are then quenched from a red heat. The surface hardness produced by this method is, however, usually limited to some 0.1 mm thick compared with the 1.0 mm or so which can be obtained by the large-scale case-hardening process.

23.6 Alloy steels

Straight carbon steels are much less employed these days, since the addition of other elements such as tungsten, nickel, chromium,

Materials

vanadium, molybdenum and cobalt enables special qualities to be imparted to the steel.

High-speed steels, which may contain 0.7 per cent carbon, 18 per cent tungsten and 5 per cent chromium, with perhaps molybdenum, vanadium or cobalt, will cut at 5 to 15 times the speed of carbon-steel tools and maintain hardness at a red heat.

During the quenching of large carbon-steel tools, cracks are sometimes produced by the contraction of the metal. This trouble can be avoided by employing an alloy tool steel, a typical example of which contains 1 per cent carbon, 1 per cent manganese, 0.75 per cent chromium, 0.5 per cent tungsten.

When exceptional strength and toughness are required, high-tensile steels are available; these usually contain principally chromium and nickel, with some molybdenum.

Carbon usually remains an essential element in all these alloy steels, and they are nearly always employed in a heat-treated condition to take full advantage of their properties.

23.7 Heat treatment of alloy steels

High-speed steels are usually hardened by raising them gradually to about 850°C, then rapidly to a white (almost melting) heat of about 1300°C, and then quenching in oil or cooling in a blast of cold air. Secondary hardening by reheating to about 500°C is often carried out after quenching to further improve the qualities of the steel.

The alloy tool steels and high-tensile steels are generally heat treated by quenching in oil from about 850°C, followed by tempering at a temperature determined by the particular use of the alloy.

For the development of the latent qualities of modern alloy steels and special cast irons, accurate heat treatment is essential, and special furnaces and pyrometers are therefore necessary. For example, many high-speed steels oxidize and scale badly when heated, and the furnace uses a controlled atmosphere from which oxygen is excluded.

23.8 Cast iron

Cast iron contains about 2.5–4 per cent carbon together with small quantities of silicon, sulphur, phosphorus and manganese. Grey cast iron is cheap, flows readily into intricate moulds, can be easily machined and, owing to presence of free carbon in the form of graphite, forms excellent bearing surfaces.

23.9 Malleable cast iron

The low tensile strength and brittleness of cast iron can be improved by packing the castings in boxes surrounded by haematite – an iron oxide – and heating for several days at about 900°C. Some of the carbon is oxidized from the castings by this process, and near the surface the material corresponds to mild steel.

23.10 Special cast irons

With the aid of carefully controlled heat treatment and alloys containing nickel, copper, chromium or molybdenum, special cast irons, such as Meehanite, are now produced which can be used for such highly stressed parts as crankshafts and camshafts.

These alloy cast irons have the advantages of cheapness, easier machining properties, greater resistance to wear, more uniform density, and an improved vibration damping capacity when compared with the steel forgings which they can replace.

23.11 Production surface hardening

Chill casting involves shaped iron 'chills' inserted in the casting mould to abruptly cool and harden the surface of the molten metal, e.g. alloy cast-iron camshafts.

For *flame hardening*, the surface of the component is quickly heated to 800–900°C by oxyacetylene jets, then quenched. In *induction hardening* the surface is heated by eddy currents induced from a surrounding coil. A brief but heavy high-frequency current – e.g. 10 000 A, 10 000 Hz – is passed before quenching.

Nitriding is a process of heat soaking suitable alloy ferrous components in ammonia gas (NH_3), e.g. for 24 hours at 500°C. *Tufftriding* is a shorter process, e.g. for 2 hours at 570°C, using sodium cyanate (NaCNO). The nitrides and carbides produced in the treated components endow them with exceptional wear resistance, e.g. for crankshaft journals.

23.12 Aluminium

Pure aluminium is a ductile and malleable light metal with a high conductivity of heat and electricity. It resists corrosion by the formation of a surface film of oxide which, if required, can be thickened and hardened by the electrolytic process of anodizing.

With some 4 per cent copper, 0.5 per cent magnesium and 0.5 per cent manganese, aluminium forms the duralumin alloys used for tubes, bars, sheets, forgings and stampings. These (and several other alloys) have the property of *age hardening* – that is, their strength and hardness increase for some days after production, so that with suitable heat treatment duralumin can have a tensile strength three times that of pure aluminium.

Where the strength must be retained at high temperatures, e.g. for pistons, other alloys such as the Y alloys, containing some 4 per cent copper, 2 per cent nickel and 1.5 per cent magnesium, are employed, usually in the heat-treated condition. More complex aluminium alloys of greater strength and demanding precise heat treatment have been developed, such as RR77, which has a tensile strength as great as carbon steel and on a weight basis is three times as strong. It contains 2.5–3.0 per cent copper, 2–4 per cent magnesium and 4–6 per cent zinc with some silicon, iron, manganese and titanium.

When used for castings, aluminium is frequently alloyed with about 12 per cent silicon, which not only increases its strength but makes it run well into moulds.

23.13 Magnesium

Magnesium is about 40 per cent lighter than aluminium, and when suitably alloyed offers one of the highest strength/weight ratios of ordinary workshop materials. Elektron crankcases etc. help to improve the power/weight ratio of CI engines.

A typical composition contains about 8 per cent aluminium, 0.5 per cent zinc and 0.25 per cent manganese, and after heat treatment has a tensile strength equal, on a weight basis, to alloy steel.

A further asset of magnesium is the ease with which it can be machined.

23.14 Copper

Copper, one of the few metals used in the pure state, is a very ductile and malleable metal with a very high conductivity of heat and electricity – exceeded only, and to a slight extent, by silver. It is resistant to corrosion, can be easily joined by soldering, brazing or welding, and forms a great many useful alloys.

23.15 Brass

When copper is alloyed with zinc, brasses are formed. Those containing less than 36 per cent zinc are widely used for cold working. Above this proportion harder and stronger brasses are obtained which are usually worked hot.

Brass can be strengthened by the addition of aluminium, tin, iron and manganese to form the high-tensile brasses.

23.16 Bronze

Bronze was originally an alloy of copper and tin, but the word has come to be used for copper alloys containing no tin. Gunmetal, frequently used for corrosion-resisting castings, contains about 10 per cent tin and 2 per cent zinc. Phosphor bronze usually has about 10 per cent tin with 0.5 per cent phosphorus. An important series of copper alloys is the aluminium bronzes, containing from 6 per cent to 11 per cent aluminium. Those with the higher proportions often have additions of iron and nickel and can have their strength considerably increased by heat treatment.

23.17 Zinc

Pure zinc is a rather brittle metal that resists corrosion, and it is used for this property in processes like galvanizing, Brylanizing, sherardizing etc.

Zinc alloys used for pressure die-casting contain some 4 per cent aluminium, 3 per cent copper and 0.02 per cent magnesium, and are therefore known as Mazak ('k' for copper).

Chapter 24

Safety in the motor vehicle workshop

The following list – from practical experience – is intended to stimulate thought on workshop safety. Should an accident occur, a knowledge of first aid and the means for speedily obtaining medical assistance may avert serious consequences.

Danger	Prevention
Floor	
Slipping on oil, grease, trolleys etc.	Floors and passages clear and clean
Pit	
Falls	Guard rails around open pit
	Boards replaced where applicable
Concentration of fumes	Adequate ventilation
Electrical equipment	Prevention of dampness
	Low-voltage equipment
Benches and racks	
Falling 'weights'	Well-ordered workbench
	Secure storage of heavy items
Jack, hoist, crane, lifting tackle	
Slipping load	Invariable use of adequate axle stands. Chocked wheels
Failure due to faults, weakened components	Regular maintenance, periodical inspection and testing
Overloading failure	Use within capacity – care with loaded commercial vehicles
Press	
Overloading failure. Uneven loading causing explosive ejection of component from press	Care in use

Danger	Prevention
Drill	
Revolving spindle or chuck entangling clothing or hair	Correctly adjusted guards. Protective clothing, tight-fitting overall, cap
Work revolving with drill (particularly on breakthrough)	Work secured by clamps or machine vice to drilling table
Grinder	
Work jamming between stone and rest	Correctly adjusted work-rest
Splinters or grinding particles – eye risk	Invariable use of goggles or drill visor
Centrifugal wheel failure	Specified installation and guard arrangements
Machines in general	
Injury from exposed moving parts	Correctly adjusted guards. Careful use
Inability to stop machine in emergency, owing to badly positioned stop button	Correct installation
Hammer	
Loose head, greasy face or handle	Correct maintenance and use
Flying chips, possible supersonic velocity with eye risk	Care. Never strike hardened steel, e.g. another hammer head. Glasses or goggles worn when descaling etc.
Striking hand holding chisel etc.	Care. Experience
Spanner	
Slipping injury	Use of correct size and type of spanner
File	
Injury by tang	Invariable use of well-fitting handle
Swarf and turnings	
Cut hands	Clear cuttings with scraper or brush
Electrical equipment	
Shock or electrocution	Correctly installed and earthed equipment, regularly maintained. Special care with portable equipment. Special care under damp conditions. Use of low-voltage equipment where applicable

Safety in the motor vehicle workshop

Danger	Prevention
	Knowledge of artificial respiration may be vital
Compressed-air equipment	
Receiver failure	Regular maintenance, drainage Periodical inspection
High-pressure air	Never 'breathe' compressed air or oxygen (see 'fluid jets' to follow)
Commercial tyre repairs	
Attempted tyre removal without complete deflation	Ensure complete deflation before attempting removal
Insecure spring rim blown off during reinflation	Cleanliness and examination of parts before refitting Inflation behind metal grille
Compressed gases for welding or cutting	
Fire risk from bottle, hose or equipment leakage	Bottles securely retained in upright position Hoses and equipment in sound condition No oil or grease on connections
Welding	
Conjunctivitis – arc-eye from ultraviolet rays	Invariable use of goggles or shield
Burns or spatter	Use of shields, protective clothing
Heat treatment	
Improper heat treatment of stressed components, e.g. fracture of straightened axle beam	Repairs strictly limited to facilities and knowledge available
Explosive gases	
Petrol/air mixture produced during petrol-tank repairs	Thorough steaming out and testing before heating fuel containers
Acetylene/oxygen/propane etc./air mixtures from welding or cutting gas cylinders	Care. Avoiding leaks Never use oxygen instead of a compressed-air supply Beware of collection of explosive, denser than air, gas in pits, etc.
Fires	
Sparks, cigarettes, hot metal or engine parts igniting combustible material or explosive gas	Well-ordered workshop, no accumulation of waste Smoking regulations. Care

Danger	Prevention
	Knowledge of the location and methods of using fire-fighting equipment
Liquid chemicals	
Cornea damage to eye from splash of ethylene-glycol antifreeze	Care
Dermatitis from diesel fuel	Avoid prolonged contact. Use neutral oil whenever possible for testing
Burns from alkali degreasing liquids or from acids	Avoid splashes. Protective clothing. Copious washing with clean water if splashed
Fluid jets	
Penetration of the body by spray or jet from:	Ensure jet or spray is always safely directed and controlled
Diesel injectors or injection test equipment	
High-pressure greasing equipment	
Compressed-air or gas nozzle	Never direct air or gas jet towards the body
Fumes and dust	
Health hazard from fumes from: Degreaser	Adequate fume and dust extraction and ventilation
Spray painting Welding, e.g. zinc alloys	Care, particularly if working under abnormal conditions
Exhaust gas, particularly carbon-monoxide poisoning	
Gas produced by fire-extinguisher fluids on red-hot metal, or by inhaling extinguisher vapour through lighted cigarette	
Dust produced during grinding operations	
Asbestos dust from brake linings	
Typical dangerous activities	
Risk of serious injury from:	
Attempting to motor test a dynamo from the revolving rear wheel of a vehicle or by an improvised belt drive from rotating machinery	

Safety in the motor vehicle workshop

Danger

Disintegration of ball or roller bearing blown-off and spun by compressed-air jet after cleaning

Inserting fingers to check alignment during reassembly of springs, gearbox, etc. if a slip occurs

Drive engagement when engine testing the vehicle with automatic – or limited-slip – transmission

Prevention

Appendix
Conversion factors

Length	1 mile	= 1.609 km
	1 foot	= 0.3048 m
	1 inch	= 25.4 mm
Area	(length conversion factors)^2	
Volume	(length conversion factors)^3	
	1 gallon UK	= 4.546 litres
	1 gallon US	= 3.785 litres
Velocity	1 mile/h	= 0.447 m/s
	1 km/h	= 0.2778 m/s
Mass	1 ton	= 1016 kg
	1 lb	= 0.4536 kg
	1 tonne	= 1000 kg
Density	1 lb/in^3	= 27 680 kg/m^3
Force	(g = 9.807 m/s^2)	
	1 tonf	= 9.964 kN
	1 lbf	= 4.448 N
	1 kgf	= 9.807 N
Torque	1 lbf ft	= 1.356 Nm
Energy	1 ft lbf	= 1.356 J
	1 calorie	= 4.187 J
	1 Btu	= 1.055 kJ
	1 Chu	= 1.899 kJ
	1 Btu/lb	= 2.326 kJ/kg
	1 Btu/ft^3	= 37.26 kJ/m^3
Power	1 hp	= 745.7 W
	1 ft lbf/s	= 1.356 W
	1 metric hp	= 735.5 W
Pressure	1 tonf/in^2	= 15.444 MPa
	1 lbf/in^2	= 6.895 kPa
	1 kgf/cm^2	= 98.067 kPa
	1 in Hg	= 3.386 kPa
	1 in H$_2$O	= 249 Pa
	1 bar	= 100 kPa

Fuel
        1 mile/gal UK = 0.354 km/litre
        1 mile/gal US = 0.425 km/litre
        1 lb/bhp h = 0.608 kg/kWh
        1 pint/bhp h = 0.762 litre/kWh

Index

Acceleration, 6
Accelerator pump, 147
Ackerman linkage, 46
Active suspension, 41
Additives, diesel fuel, 163; oil 111; petrol 133
Advance, ignition, 189
Air, cooling, 122; filter, 137; fuel ratio, 142; silencing, 137
Alcohol, 133, 134
Alignment, chassis, 35; wheel, 49
Alloy steel, 300; heat treatment, 301
Alternator, 201
Aluminium, 302
Ammeter, 182
Angle, camber, 47; castor, 48; dwell, 193; slip, 49
Annealing, steel, 299
Antifreeze, 129
Anti-roll bar, 44
Atmospheric pressure, 14
Atomizer, 175
Automatic, starting devices, 155; transmission, 240
Axle, beam, 51; casing, 258; dead, 268; rear, 258; two-speed, 249

Balance, clutch, 214; engine, 78; propeller shaft, 245; wheel, 64
Ballasted ignition, 189
Ball bearing, 92
Battery, 195; maintenance, 196
Baulk ring, 227
Bearing metals, 93 (table)
Belt, generator, 206; toothed, 99, 100
Benzole, 133
BMEP-IMEP, 73 (graph)

Body, 31; repair, 33
Brake, actuation, 272, 277; adjustment, 278; auxiliary, 293; compensation, 273; defects, 295; disc, 284; dive, 270; divided system, 286; efficiency, 272; fade, 271; hydraulic, 279; inboard, 286; layout, 286; maintenance, 294; mechanical, 277; operation, 273; power-operated, 292; principles, 269; servo-operated, 288; self-energizing, 276; shoe, 271, 273
Brass, 303
Bronze, 304

Cables, 183
Calibration, injection pump, 169
Calorific value, 133
Camber angle, 47
Camshaft, 99
Capacity, battery, 196; engine, 75
Capacitor, 187, 188; discharge ignition, 194
Carbon steel tempering colours, 300 (table)
Carburation, 140; emission control, 154; faults, 153
Carburetter, constant vacuum, 147; fixed choke, 141; simple, 140
Case hardening, 300
Cast iron, 301
Castor angle, 48
Celsius, 14
Centre of gravity, mass, 10
Centre point steering, 47
Centrifugal force, 12
Cetane number, 162

Index

Chapman strut, 267
Charging system, 195
Chassis, 31; repair, 33
Clutch, 207; defects, 214; diaphragm, 211; maintenance, 213; withdrawal mechanism, 209
Coefficient of, friction, 13, 269; linear expansion, 19 (table)
Coil ignition, 186; maintenance, 192
Colours of steel, 298 (table)
Combustion, diesel, 162; petrol, 76, 132
Combustion chamber, 79, 153
Compensation, brake, 273; carburation, 141
Compression gauge assessment, 107 (table)
Compression, pressure, 106; ratio, 75
Compression-ignition, operation, 160; maintenance, 178
Condenser, 187, 188
Conduction, electrical, 183; thermal, 17, 18
Connecting rod, 95; alignment, 96
Constant velocity joint, 241
Construction, integral, 31, 32
Conversion factors, 310
Cooling system, air, 122; directed, 126; sealed, 127; maintenance, 130
Copper, 303
Cornering force and power, 51 (graph)
Couple, 11
Crankcase, 89; emission, 156
Crankshaft, 91; layout, 78
Crown wheel, 247; tooth-contact, 253
Cut-out, 200
Cycle, four-stroke, 68; two-stroke, 69; CI, 160
Cylinder block, 89

Damper, hydraulic, 42; vibration, 91
De Dion tube, 268
Delay period, 162, 175
Delivery, maximum, valve, 169
Density, 5
Detonation, 76, 133
Diaphragm spring clutch, 211
Differential, 247, 269; applications, 251

Direct injection, 163
Distributor, 188; type pump, 170
Drawing, 27
Drift angle, 49
Drive, final, 247; front and rear wheel, 217; ratchet, 251; shafts, 245
Driving-axle torque, 219 (graph)
Dual characteristics, 142
Duralumin, 303
Dwell angle, 193
Dynamo, 198

Earth return, 184
Efficiency, brake, 272; mechanical, 72; thermal, 74; volumetric, 73
Electrical system, 180
Emission control, 153
Energy, conversion, 18, 74; distribution, 74
Engine, comparison, 77; mounting, 32; position, 34
Epicyclic, gearing, 238; overdrive, 230; two-speed final drive, 249
Exhauster emission, 154 (graph)
Exhauster pump, 289
Expansion, 18; coefficient, 19

Fan, 122, 123; control, 129
Ferrous metals, 298
Filtration, air, 136; diesel fuel, 165; oil, 117
Final drive, 247; epicyclic, 249; maintenance, 252; ratio, 218; worm, 248
Firing, interval, 77; order, 78
Fleming's rule, left-hand, 202; right-hand, 197
Fluid-coupling slip, 236 (graph)
Flywheel, engine, 92; fluid, 235; vibration damper, 92
Force, centrifugal, 12; cornering, 50; friction, 12; gravity, 6; inertia, 11; retarding, 269
Four-stroke cycle, 68
Freewheel, transmission, 230; starter drive, 205
Friction, 12; brake shoe, 269
Front wheel assembly, 59
Frost protection, 129

313

Fuel, petrol, 132; diesel, 162
Fuse, 185

Gas, exhaust, 132; law, 15; pollutant, 152
Gasket, 105
Gauge, pressure, vacuum, 106
Gear-box, all-indirect, 231; auxiliary, 229; constant-mesh, 225; defects, 232; epicyclic, 238; five-speed, 229; lubrication, 232; overdrive, 230; ratios, 218; sliding-mesh, 219; synchromesh, 227; two-speed transfer, 231
Gear ratio, 217, 221, 222; additional, 228; changing, 226; epicyclic, 238
General gas law, 15
Generator, 197
Geometry, 3; steering, 46; suspension, 44
Governor, air velocity, 149; hydraulic, 174; mechanical, 173
Gravity, 6; centre of, 10
Grease, 120
Gudgeon pin, 95
Gunmetal, 304

Halogen bulb, 185
Handbrake, 282
Hardening, case, 300; production surface, 302
Headlamp, 185
Heat, latent, 17; quantity, specific, 16; transfer, 17; unit, 16
Heat treatment, 301
Heavy discharge tester, 196
Heron combustion chamber, 80
Hotchkiss drive, 263
Hot spot, 150
Hub, front, 59; rear, 258
Hydraulic operation, brake, 279; clutch, 210
Hydrocarbon, emission, 153; fuel, 132
Hydrometer, antifreeze, 130; battery, 196
Hypoid drive, 247

Idling, devices, 146; hot, 155
IFS, 52; double-transverse link, 53; MacPherson, 54

Ignition system, ballasted, 189; battery-coil, 186; electronic, 193; maintenance, 192; timing, 189, 191
Indicated power, pressure, 70
Indicator diagram, 70, 71 (graph)
Induction, electro-magnetic, 186
Inertia, 5, 11; drive, 204
Injection, direct, 163; indirect, 165; pump, 166, 167, 170
Injection system, diesel, 165, 170; petrol, 157
Injector, 175; testing, 177
Intake air temperature, 154
Interconnected suspension, 40
IP-BP, 74 (graph)
Iron, cast, 301; malleable, special, 302
IRS, 267; final drive, 261

Joint, constant velocity, 241; cylinder head, 105; flexible, 242; Hooke's, 241

Kelvin, 14
KPI, 47

Latent heat, 17
Law, gas, 15; Hooke's, 36; Ohm's 180
Layout, brake, 286; conventional, 78; engine, 36
Liner, cylinder, 89
Lining, brake, 271
Live axle, 258
Lubrication system, 110; dry-sump, 116; force-feed, 110; maintenance, 120; pressure, 119; splash, 111

Magnesium, 303
Magnetic field, 186
Manifold, 150
Masked inlet valve, 163
Mass, 5
Master cylinder, 280, 281; tandem, 286
May combustion chamber, 81
Mazak, 32
Mechanical efficiency, 72
Mensuration, 2
Metals, ferrous, 298; non-ferrous, 302
Melting points, 15 (table)
Mixture strength, 152

Index

Moments, 9
Morse test, 87
Mountings, engine, 32
Multicylinder engine, 77
Multiplying factors, metric, 1

Normalizing, steel, 300
Nitriding, 302

Octane number, 76, 133
Offset, steering, 47
Ohm's law, 180
Oil, changing, 120; cooling, 119; filtration, 117; pressure, 119; pump, 113; sealing, 116; viscosity, 110
Open chamber injection, 163
Orthographic projection, 27
Overdrive, 230
Overheating, 130
Oversteer, 50

Panhard rod, 265
Petrol, 132; ratings, 134 (table)
Phasing, injection pump, 169
Piston, 96; clearance, 97; ring, 98; speed, 12
Polarity, 189
Power, 8, 217; brake, 72; cornering, 50; indicated, 70; requirement, 219; –weight ratio, 77
Pre-engaged starter motor, 204
Pre-ignition, 77
Preload, bearing, 252
Pressure, 13; absolute, atmospheric, 14; brake, 72; gauge, 14; indicated, 70; radiator cap, 126
Progression hole, 146
Propeller shaft, 245
Pump, injection, 166; oil, 113; petrol, 134, 135; water, 125
Pythagorus' theorem, 3

Quantity of heat, 16

Radiator, 124; pressurized, 125
Ratio, compression, 75; epicyclic, 238; gearbox, final drive, 218
Rear axle, 258; control, 265; torque curves, 218
Rear drive, live, 258; sprung, 261

Regulator, dynamo, 198
Relative density, 5 (table)
Remote control, gearchange, 223
Resistance, parallel, 182; series, 180; tractive, 216
Resistivity, 182
Regard, ignition, 189; effect, 191
Rim, commercial, 64; well-base, 63
Rockers, valve, 100
Roll, body, 44
Roller bearing, 92

SAE number, 110
Safety, workshop, 305
Selector, 222; interlock, single-rail, 224
Self-levelling suspension, 42
Semiconductor, 201
Servo-assisted, braking, 288; steering, 58
Shaft, drive, propeller, 245
SI, 1
Sliding-mesh gearbox, 220
Slip, angle, 49; fluid flywheel, 238
Solenoid, 186; switch, 204, 206
Sparking plug, 191
Specific heat capacity, 16 (table)
Speed, 6; piston, 12
Spring, air, 39, 40; characteristics, 37 (graph); coil, 39; leaf, 36; rubber, 39; tapered leaf, 38; valve, 105; variable rate, 39
Starter motor, 203; maintenance, 205
Starting devices, 143; automatic, 155; CI, 174
Static timing, 191
Steel, 298; heat treatment, 300
Steering, box, 55; centre-point, 47; linkage, 58
Strangler, 143
Stress, 14
Stroboscopic timing, 191
Strut, Chapman, 267; MacPherson, 44
Suspension, 36; active, 41; geometry, 44; IFS, 52; interconnected, self-levelling, 40; IRS, 267
Synchromesh, 227

Tappet, 100; hydraulic, 104
Temperature, 14

315

Tempering colours, 300
Thermal conductivity, 18 (table)
Thermostat, 127
Timing, ignition, 191; valve, 68
Toe-in, 49; adjustment, 59
Torque, 91; converter, 236; driving, 218, 219; engine, 73, 74; fluctuation, 77; reaction, 32, 33, 263; tube, 264
Torque converter ratio efficiency, 238 (graph)
Track adjustment, 59
Tractive effort, 216; resistance, 216 (graph)
Transaxle, 262
Trigonometry, 4
Tube, De Dion, 268; torque, 264; tyre, 62
Turbocharger, 179
Turbulence, CI, 163; petrol, 79, 154
Two-stroke cycle, 69
Tyre, 61; balance, 64; fitting, 63; nomenclature, 62; usage, 63; wear, 65

Understeer, 50

Units, conversion, 310; heat, 16; SI, 1
Universal joint, 241

Vacuum gauge, 108; assessment, 108 (table); servo, 288
Valve, clearance, 100; engine, 103; spring, 105; timing, 68; tyre, 62
Viscosity, oil, 110; index, 111
Volatility, 133
Voltmeter, 182

Water pump, 125
Watt's linkage, 265
Wave winding, 204
Wedge chamber, 80
Wet liner, 90
Weight, 5; transfer, 218, 270; unsprung, 36
Wheel, alignment, 49; assembly, 59; balance, 64; offset, 47; tolerance, 64
Wiring, circuit, 184; diagram, 185
Work, 8, 9
Worm and, roller, 57; wheel, 248

Zinc, 304